SpringerBriefs in Mol

For further volumes:
http://www.springer.com/series/8898

Zhigang Shuai · Linjun Wang
Chenchen Song

Theory of Charge Transport in Carbon Electronic Materials

Zhigang Shuai
Department of Chemistry
Tsinghua University
100084 Beijing
People's Republic of China
e-mail: zgshuai@tsinghua.edu.cn

Chenchen Song
Department of Chemistry
Tsinghua University
100084 Beijing
People's Republic of China
e-mail: Scc.007@163.com

Linjun Wang
Service de Chimie des Matériaux
Nouveaux
Université de Mons
7000 Mons, Belgium
e-mail: linjun.wang@umons.ac.be

ISSN 2191-5407
e-ISSN 2191-5415
ISBN 978-3-642-25075-0
e-ISBN 978-3-642-25076-7
DOI 10.1007/978-3-642-25076-7
Springer Heidelberg Dordrecht London New York

Library of Congress Control Number: 2011941672

Printed on acid-free paper

Springer is part of Springer Science+Business Media (www.springer.com)

Preface

Organic electronics has been extensively studied for over 50 years, and is now still a rapidly developing area. Charge carrier mobility is at the center of these electronic devices. This book describes recent progresses in developing computational tools to assess the intrinsic carrier mobility for organic and carbon materials at the first-principles level. According to the electron–phonon coupling strength, we classify the charge transport mechanism into three different categories, namely, the localized hopping model, the extended band model, and the polaron model. For each of them, we develop a corresponding theoretical approach and implement it to typical examples. A lot of successes have been achieved and the outlook is given. The authors are deeply grateful to the following collaborators: Dr. Mengqiu Long and Dr. Ling Tang, postdoctoral fellows in Shuai's group, working on developing deformation potential theory applied to organic and carbon materials; Dr. Shiwei Yin, Dr. Guangjun Nan and Dr. Xiaodi Yang, former PhD students in Shuai's group, working on developing intermolecular coupling quantum chemistry method, random walk simulation, and quantum nuclear tunneling effects within hopping scheme, respectively. The financial supports come from the National Natural Science Foundation of China, the Ministry of Science and Technology of China, the Chinese Academy of Sciences, the European Union sixth Framework, and Solvay. We are glad to receive any helpful suggestions and comments from the readers.

Beijing, September 2011 Z. G. Shuai

Contents

Chapter 1
Introduction

The charge mobility, μ, which characterizes the ability of a charge to move in a bulk semiconductor, is the essential parameter in determining the overall performance of electronic devices [1]. By definition, it is the charge drift velocity, v, acquired per driving electric field, F, i.e., $\mu = v/F$, usually expressed in unit of cm^2/Vs. In the absence of scattering, the field-induced momentum gain for an electron, $\Delta q = -eFt$, should increase linearly with the time period t. However, according to the classical Boltzmann transport picture, due to the scattering with impurities, defects, and lattice vibrations, the electron momentum is restored to its original value after the mean scattering time, τ, i.e., the average time between two consecutive scattering events. Therefore, in the steady current condition, the acquired momentum is a finite value of $\Delta q = -eF\tau$. If the charge has an effective mass m^*, the velocity becomes $v = -e\tau F/m^*$, and thus the prefactor is the mobility, $\mu = -e\tau/m^*$. Traditional inorganic semiconductors, especially silicon single crystals, possess a room-temperature mobility ranging from a few hundreds to a few thousands cm^2/Vs [2]. Such large mobility, as well as natural abundance and stability, makes silicon the most prominent electronic material. New carbon materials, such as single-walled carbon nanotubes and graphene sheets, are very promising for the next generation of electronics because their intrinsic mobility can reach up to a few hundred thousand cm^2/Vs [3]. Generally, organic materials have much lower mobility and poorer stability, and thus they are not destined to replace silicon. However, they can play an important role in next-generation electronic applications, such as large area and flexible devices, due to their processability and flexibility, among other advantages [4]. Designing new organic materials with large mobility has been a formidable task in the past two decades and now a variety of new molecular materials have been synthesized with room temperature mobilities between 1 and 10 cm^2/Vs in thin films [5] and even larger values in single crystals [6].

On one hand, the charge transport in ideal molecular crystals has been a subject of theoretical interests for almost sixty years. In 1959, Holstein proposed the small

Z. Shuai et al., *Theory of Charge Transport in Carbon Electronic Materials*,
SpringerBriefs in Molecular Science, DOI: 10.1007/978-3-642-25076-7_1,

polaron model, which depicted a general scheme for studying charge transport in organic solids [7]. However, there are at least three reasons for the recent renaissance in theoretical interest: (1) in recent years, there have been great advances made both in synthesis of better performed molecules and in materials processing for good single crystals; (2) tremendous advances in electronic structure theory and computational technology have allowed a quantitative description for the electronic properties, e.g., intermolecular electronic couplings and electron–phonon couplings, and thus charge mobility in molecular crystals; and (3) despite the numerous theoretical studies made in the past, the comparison between experiment and theory had been always extremely difficult due to the lack of quantitative calculations as well as ultrapure single crystals in the absence of structural disorder and chemical impurity. Thus, it is difficult to judge the applicability of the different levels of approximation to solve the Holstein model Hamiltonian. These facts have hindered our understanding of the intrinsic transport mechanism. On the other hand, the charge transport in real organic devices has often been described by phenomenological disorder models. The most successful one was developed by Bässler and coworkers, assuming Gaussian random site energies with uncorrelated static disorder and that charge hops between sites through absorbing or emitting phonons [8]. This model was later modified using a correlated energetic disorder to account for more realistic situation with charge–dipole interaction [9]. With such model, both temperature and field dependence for the mobility in organic devices have been successfully described.

Nevertheless, from the material design point of view, a microscopic model for the charge transport property is highly desirable. Approximately, different scattering relaxation times arising from optical phonons, acoustic phonons, defects, and impurities can be added up as: $1/\tau = 1/\tau_{op} + 1/\tau_{ac} + 1/\tau_{def} + 1/\tau_{imp} + \ldots$, and thus the shortest relaxation time contributes the most to the overall mobility. Since defects and impurities are extrinsic features that can be minimized through molecular design and material processing, strong interest has been given to determine the intrinsic mobility arising from scattering with phonons for a given material. There are two widely used transport pictures: the band model for delocalized electrons [10] and the hopping model for localized charge [11]. In oligoacenes and rubrene single crystals, experimental evidence has shown that a band picture is more appropriate. In recent years, significant progresses have been achieved in a number of areas, for example: (i) through first-principles calculations, Brédas and coworkers have systematically explored the molecular parameters relevant to charge transport based on the semiclassical Marcus electron transfer theory [12], e.g., the intermolecular electronic couplings, the molecular reorganization energy, and the polarization energy in the bulk [13]; (ii) Munn and Silbey developed a variational approach to solve the Holstein model and compared the local and nonlocal electron–phonon coupling contributions to band and hopping transport [14]—they concluded that the nonlocal terms tend to enhance the hopping behavior; (iii) Kenkre and coworkers derived a working expression for mobility and presented a unified quantitative explanation of the temperature dependence of mobility in naphthalene crystal by assuming a direction-dependent local electron–phonon coupling constant [15]; (iv) Bobbert and coworkers generalized the Holstein model to the Holstein-Peierls model by including the nonlocal electron–phonon coupling terms while keeping molecule rigid [16]. We later extended such model to incorporate both inter- and intra-molecular vibration modes, with all the parameters evaluated by DFT [17]; and (v) a time-dependent Schrödinger equation with a Su–Schrieffer–Heeger Hamiltonian has been solved numerically for the charge diffusion process with local electron–phonon coupling by Hultell and Stafström [18] and with nonlocal electron–phonon coupling by Troisi and Orlandi [19], where phonons are both treated classically.

In this book, we present our recent progresses toward better understanding of charge transport in organic materials and quantitative predictions of carrier mobility through first-principles calculations. We classify the charge transport mechanism into three categories according to electron–phonon coupling strength, for each of which we have developed a corresponding theoretical method. The purpose is to develop a computational tool for assessing the intrinsic charge mobility of the organic and carbon materials. In the first category, the electron interacts strongly with intramolecular vibrations, namely, the intermolecular electron coupling is much less than the molecular reorganization energy. In this case, the electron is fully self-localized, i.e., each molecule acts as a trap, and the charge transport can be viewed as an intermolecular hopping process [11, 20–26]. It is appropriate to apply the Marcus theory to calculate the intermolecular charge transfer rates, and a more elaborate and appropriate description is to incorporate nuclear tunneling effects to account the quantum nature of molecular vibration, which we found is essential since the intramolecular vibration is generally of high frequency, invalidating the classical charge transfer theory [25]. The charge diffusion can be modeled through a random walk numerical simulation, much as in the phenomenological disorder model. Since the molecular parameters enter explicitly into evaluation of the charge diffusion coefficient, this approach can be used for molecular design toward high charge mobility. In this picture, it is found that the dynamic disorder is very much dependent on the space dimension, and sometimes, it leads to the phonon-assisted current, namely, dynamic disorder enhances the charge mobility [26]. Applications have been performed to a variety of molecular materials, e.g., siloles [20], triphenylamines [21, 22], oligothiophenes [24], and oligoacenes [25, 26]. In the second category, the transfer integral is comparable to the reorganization energy, and the localized picture should not be imposed at first-hand. We consider the Holstein-Peierls polaron model, where both intra- and inter-molecular vibrations are considered to be locally and nonlocally coupled with the electron. The vibrations are treated as optical phonons [16]. All the parameters in the Hamiltonian are evaluated through DFT calculations, allowing us to establish quantitative structure–property relationship and prediction. The approach is applied to naphthalene crystal [17, 27, 28]. Both the temperature and pressure dependence of mobility have been systematically studied, considering the thermal expansion and press compression of the lattice [28]. In the third category, the coherent length of electrons is assumed to be very long, matching the wavelength of acoustic phonons, much as in inorganic semiconductors. In this case, the scattering process is modeled by a deformation potential formalism, which is the band model including only the lattice scatterings by the acoustic deformation potential. With the mobility formula based on the Boltzmann transport equation and the effective mass approximation, the role of acoustic phonons in naphthalene crystal is reexamined with DFT calculations [29]. Typical examples are also shown for the graphene [30] and graphdiyne [31] sheets and nanoribbons.

References

1. V. Coropceanu, J. Cornil, D.A. da Silva Filho, Y. Olivier, R.J. Silbey, J.-L. Brédas, Chem. Rev. **107**, 926 (2007)
2. W.F. Beadle, J.C.C. Tsai, R.D. Plummer (eds.), *Quick Reference Manual of Semiconductor Engineers* (Wiley, New York, 1985)
3. C. Berger, Z. Song, X. Li, X. Wu, N. Brown, C. Naud, D. Mayou, T. Li, J. Hass, A.N. Marchenkov, E.H. Conrad, P.N. First, W.A. de Heer, Science **312**, 1191 (2006)
4. A.J. Heeger, N.S. Sariciftci, E.B. Namdas, *Semiconducting and Metallic Polymers* (Oxford University Press, New York, 2010)
5. C.R. Newman, C.D. Frisbie, D.A. da Silva Filho, J.L. Brédas, P.C. Ewbank, K.R. Mann, Chem. Mater. **16**, 4436 (2004)
6. V.C. Sundar, J. Zaumseil, V. Podzorov, E. Menard, R.L. Willett, T. Someya, M.E. Gershenson, J.A. Rogers, Science **303**, 1644 (2004)
7. T. Holstein, Ann. Phys. (N.Y.) **8**, 325 (1959)
8. H. Bässler, Philos. Mag. B **50**, 347 (1984)
9. A. Dieckmann, H. Bässler, P.M. Borsenberger, J. Chem. Phys. **99**, 8136 (1993)
10. M.E. Gershenson, V. Podzorov, A.F. Morpurgo, Rev. Mod. Phys. **78**, 973 (2006)
11. L.J. Wang, G.J. Nan, X.D. Yang, Q. Peng, Q.K. Li, Z.G. Shuai, Chem. Soc. Rev. **39**, 423 (2010)
12. J.L. Brédas, D. Beljonne, V. Coropceanu, J. Cornil, Chem. Rev. **104**, 4971 (2004)
13. J.E. Norton, J.L. Brédas, J. Am. Chem. Soc. **130**, 12377 (2008)
14. R. Silbey, R.W. Munn, J. Chem. Phys. **72**, 2763 (1980)
15. V.M. Kenkre, J.D. Anderson, D.H. Dunlap, C.B. Duck, Phys. Rev. Lett. **62**, 1165 (1989)
16. K. Hannewald, V.M. Stojanović, J.M.T. Schellekens, P.A. Bobbert, G. Kresse, J. Hafner, Phys. Rev. B **69**, 075211 (2004)
17. L.J. Wang, Q. Peng, Q.K. Li, Z.G. Shuai, J. Chem. Phys. **127**, 044506 (2007)
18. M. Hultell, S. Stafström, Chem. Phys. Lett. **428**, 446 (2006)
19. A. Troisi, G. Orlandi, Phys. Rev. Lett. **96**, 086601 (2006)
20. S.W. Yin, Y.P. Yi, Q.X. Li, G. Yu, Y.Q. Liu, Z.G. Shuai, J. Phys. Chem. A **110**, 7138 (2006)
21. Y.B. Song, C.A. Di, X.D. Yang, S.P. Li, W. Xu, Y.Q. Liu, L.M. Yang, Z.G. Shuai, D.Q. Zhang, D.B. Zhu, J. Am. Chem. Soc. **128**, 15940 (2006)
22. X.D. Yang, Q.K. Li, Z.G. Shuai, Nanotech. **18**, 424029 (2007)
23. L.Q. Li, H.X. Li, X.D. Yang, W.P. Hu, Y.B. Song, Z.G. Shuai, W. Xu, Y.Q. Liu, D.B. Zhu, Adv. Mater. **19**, 2613 (2007)
24. X.D. Yang, L.J. Wang, C.L. Wang, W. Long, Z.G. Shuai, Chem. Mater. **20**, 3205 (2008)
25. G.J. Nan, X.D. Yang, L.J. Wang, Z.G. Shuai, Y. Zhao, Phys. Rev. B **79**, 115203 (2009)
26. L.J. Wang, Q.K. Li, Z.G. Shuai, L.P. Chen, Q. Shi, Phys. Chem. Chem. Phys. **12**, 3309 (2010)
27. L.J. Wang, Q.K. Li, Z.G. Shuai, J. Mol. Sci. (Chinese) **24**, 133 (2008)
28. L.J. Wang, Q.K. Li, Z.G. Shuai, J. Chem. Phys. **128**, 194706 (2008)

29. L. Tang, M.Q. Long, D. Wang, Z.G. Shuai, Sci. China Ser. B-Chem. **52**, 1646 (2009)
30. M.Q. Long, L. Tang, D. Wang, L.J. Wang, Z.G. Shuai, J. Am. Chem. Soc. **131**, 224704 (2009)
31. M.Q. Long, L. Tang, D. Wang, Y.L. Li, Z.G. Shuai, ACS Nano **5**, 2593 (2011)

Chapter 2
Hopping Mechanism

Abstract In the limit of strong electron–phonon coupling and weak intermolecular electronic coupling, a charged molecule undergoes a large geometry relaxation, which eventually traps the charge. In this case, the charge transport can be viewed as an intermolecular hopping process. With the known electron transfer rates between neighboring molecules, the charge carrier mobility can be evaluated through the Einstein relation from random walk simulations. In general, the classical Marcus electron transfer theory, which works well in the high-temperature limit. For a better understanding, we incorporate the nuclear tunneling effect arising from the intramolecular high-frequency vibrations to characterize the transport behavior at room temperature. Dynamic disorder effect arising from the intermolecular low-frequency vibrations is found to be very much materials structure or space-dimension dependent, which may give rise to the phonon-assisted current.

Keywords Hopping mechanism · Marcus electron transfer rate · Random walk simulation · Temperature dependence of mobility · Nuclear tunneling · Dynamic disorder

In Sect. 2.1, we describe the general methodology to simulate the hopping mobility using the electron transfer rate formalism. This approach is applied to different organic materials and discussed in Sects. 2.2, 2.3, 2.4. Section 2.5 investigates the nuclear tunneling effect of the intramolecular vibrations, and finally Sect. 2.7 is about the dynamic disorder effect of the intermolecular modes. For each of these improvements, application examples are presented in Sects. 2.6 and 2.8, respectively.

Z. Shuai et al., *Theory of Charge Transport in Carbon Electronic Materials*,
SpringerBriefs in Molecular Science, DOI: 10.1007/978-3-642-25076-7_2,

2.1 General Methodology

This chapter deals with strong electron–phonon interaction limit that the charge is regarded as localized in a single molecule. The charge transport consists of successive hopping from molecule to molecule, overcoming the trap caused electron scatterings with intramolecular vibrations. In the hopping picture to evaluate the charge mobility, there are two important rate processes at different spatial scales, namely, the electron transfer within molecular dimers and the electron diffusion in organic solids. The previous process can be characterized by the intermolecular electron transfer rate at the atomistic level, while the latter can be simulated at the molecular level by the random walk technique, regarding each molecule as a lattice site. When the intermolecular electronic coupling, also called electron transfer integral, V, is much smaller than the reorganization energy of the electron transfer process between a molecular dimer, λ, the electron transfer rate at high temperature, T, falls well into the hopping regime. For organic materials, the intermolecular interaction is of van der Waals type. In general, V is smaller than λ, and thus the Marcus rate is usually adopted to evaluate the room-temperature mobilities.

2.1.1 Marcus Electron Transfer Rate

The Marcus formula for electron transfer rate reads [1]

$$k = \frac{V^2}{\hbar}\sqrt{\frac{\pi}{\lambda k_B T}}\exp\left\{-\frac{(\Delta G^0 + \lambda)^2}{4\lambda k_B T}\right\} \tag{2.1}$$

Here, $\hbar$ is the reduced Planck constant, k_B is the Boltzmann constant, and ΔG^0 is the free energy difference between the initial and final molecular sites. For molecular crystals with only one type of molecules, ΔG^0 is generally zero since all molecules in the crystal are equivalent. In Eq. 2.1, V and λ are the two most important parameters, both of which are related to the material itself. Various approaches have been proposed to calculate these two parameters in the literature, as presented below.

2.1.2 Transfer Integral

The intermolecular transfer integrals can be calculated through various numerical methods. Three of them, including the direct method [2], the site-energy correction method [3], and the band-fitting method [4], have been widely used in the literature, and thus will be detailedly discussed below.

2.1.2.1 Direct Method

The direct scheme to obtain the intermolecular transfer integral was proposed by Fujita et al. in modeling scanning tunneling microscopy [2] and later adopted by Troisi and Orlandi to study the charge transport in DNA and pentacene crystal [5]. At the Hartree-Fock (HF) level, the transfer integral reads:

$$
\begin{aligned}
V_{e/h} &= \left\langle \phi^{0,\text{site1}}_{\text{LUMO/HOMO}} \middle| F^0 \middle| \phi^{0,\text{site2}}_{\text{LUMO/HOMO}} \right\rangle \\
&= \left\langle \phi^{0,\text{site1}}_{\text{LUMO/HOMO}} \middle| h_{\text{core}} \middle| \phi^{0,\text{site2}}_{\text{LUMO/HOMO}} \right\rangle \\
&\quad + \sum_{l(\text{occ.})} \left(\left\langle \phi^{0,\text{site1}}_{\text{LUMO/HOMO}} \phi^0_l \,\middle|\, \phi^{0,\text{site2}}_{\text{LUMO/HOMO}} \phi^0_l \right\rangle \right. \\
&\quad \left. - \left\langle \phi^{0,\text{site1}}_{\text{LUMO/HOMO}} \phi^{0,\text{site2}}_{\text{LUMO/HOMO}} \,\middle|\, \phi^0_l \phi^0_l \right\rangle \right)
\end{aligned}
\tag{2.2}
$$

Here, $\phi^{0,\text{site1}}_{\text{LUMO/HOMO}}$ and $\phi^{0,\text{site2}}_{\text{LUMO/HOMO}}$ represent the highest occupied molecular orbital (HOMO) and lowest unoccupied molecular orbitals (LUMO) of the two adjacent molecules when no intermolecular interaction is present. F^0 is the Fock operator for the dimer, which is calculated with the unperturbed molecular orbitals.

It has been shown that the HF bandwidth for a polymer is always about 20–30% larger than the experimental results [6]. Moreover the coupling calculated from density functional theory (DFT) is usually about 20% less than that from HF [7]. Therefore, in Eq. 2.2, it is better to adopt the Kohn-Sham–Fock operator instead of the Fock operator:

$$
F^0 = SC\varepsilon C^{-1} \tag{2.3}
$$

Here, S is the intermolecular overlap matrix, and C and ε are the Kohn-Sham molecular orbital coefficients and eigenenergies of the non-interacting dimer, respectively. In practice, the molecular orbitals of the two individual molecules are calculated separately by the standard self-consistent field procedure. These non-interacting orbitals are then used to construct the Kohn–Fock matrix. After one-step diagonalization without iteration, the orbital coefficients and eigenenergies, and thus the Fock operator for the non-interacting dimer can be calculated.

2.1.2.2 Site-Energy Correction Method

When the self-consistent field is performed to construct the Kohn-Sham–Fock operator in Eq. 2.3, Valeev and coauthors noticed that the site-energy difference should be taken into account when the dimer is not cofacially stacked [3]. For example, assuming that HOMO and HOMO-1 of the dimer result only from the interaction of the monomer HOMOs, $\{\Psi_i\}$, the dimer molecular orbitals and the corresponding orbital energies follow the secular equation:

$$
\mathbf{HC} - E\mathbf{SC} = 0 \tag{2.4}
$$

Here, **H** and **S** are the Hamiltonian and overlap matrices of the system:

$$\mathbf{H} = \begin{pmatrix} e_1 & J_{12} \\ J_{12} & e_2 \end{pmatrix} \quad (2.5)$$

$$\mathbf{S} = \begin{pmatrix} 1 & S_{12} \\ S_{12} & 1 \end{pmatrix} \quad (2.6)$$

The matrix elements above are

$$e_i = \langle \Psi_i | \hat{H} | \Psi_i \rangle \quad (2.7)$$

$$J_{ij} = \langle \Psi_i | \hat{H} | \Psi_j \rangle \quad (2.8)$$

$$S_{ij} = \langle \Psi_i | \Psi_j \rangle \quad (2.9)$$

Since the basis sets, namely the HOMOs of the adjacent individual molecules, are not orthogonal with each other, a Lowdin's symmetric transformation can be applied to get an orthonormal basis set, which also maintains as much as possible the initial local character of the monomer orbitals. With such a symmetrically orthonormalized basis, we have the effective Hamiltonian:

$$\mathbf{H}^{\mathrm{eff}} = \begin{pmatrix} e_1^{\mathrm{eff}} & J_{12}^{\mathrm{eff}} \\ J_{12}^{\mathrm{eff}} & e_2^{\mathrm{eff}} \end{pmatrix} \quad (2.10)$$

Here, the off-diagonal term is the effective transfer integral which has considered the site-energy correction: [3]

$$J_{12}^{\mathrm{eff}} = \frac{J_{12} - \frac{1}{2}(e_1 + e_2)S_{12}}{1 - S_{12}^2}. \quad (2.11)$$

2.1.2.3 Band-Fitting Method

Within the tight binding model, if the site energies of all molecules are equivalent, the energy band can be expressed as:

$$\varepsilon(\mathbf{k}) = \varepsilon_0 + \sum \varepsilon_{ij} e^{-i\mathbf{k}\cdot\mathbf{R}_{ij}} \quad (2.12)$$

Here, i is any molecule in the unit cell which has been chosen as a reference, j runs over all the chosen neighbors of molecule i, $\mathbf{k}$ is the wavevector, and $\mathbf{R}_{ij}$ is the spatial vector between molecules j and i. A first-principles density functional theory band structure can be projected to Eq. 2.12 through fitting all the transfer integrals for the corresponding molecular dimers [4].

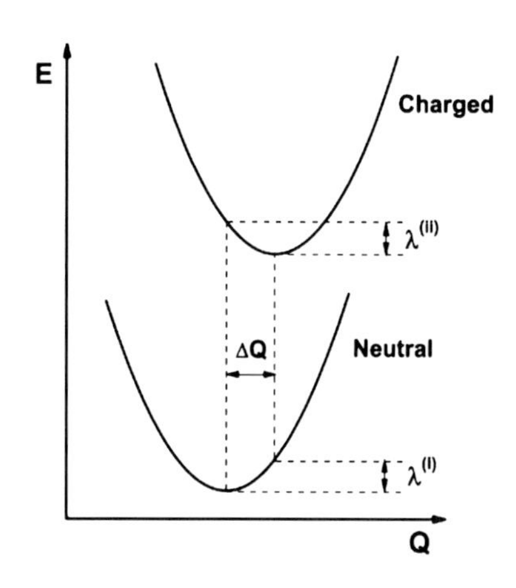

Fig. 2.1 Schematic representation of the potential energy surfaces of the neutral and charged molecules. Q is the reaction coordinate, and the sum of the two relaxation energies $\lambda^{(i)}$ and $\lambda^{(ii)}$ is the internal reorganization energy. Reproduced from Ref. [14] by permission of The Royal Society of Chemistry

2.1.3 Reorganization Energy

The reorganization energy is composed of two parts, the internal reorganization (λ_i) and the external polarization (λ_e) [8]. The former term λ_i reflects the change in molecular geometry associated with going from the neutral state to the ionized state, and vice versa. And the latter term λ_e describes the change in electronic polarization of the surrounding molecules. The external contribution is difficult to evaluate theoretically, and thus is normally neglected during discussion. In some cases, a magnitude of 0.2–0.6 eV is assumed for λ_e as will be seen in Sect. 2.2 for general understanding of the role of λ_e. In the following, we will only discuss the calculation of the internal reorganization energy at the first principle level.

2.1.3.1 Diabatic Potential Surface

If we obtain the diabatic potential surfaces of the neutral and charged molecules, as shown in Fig. 2.1, we can easily calculate the reorganization energy of the charge transfer reaction between a molecular dimer, which is a sum of two relaxation energy terms: (i) the difference between the energy of the neutral molecule in its equilibrium geometry and in the relaxed geometry characteristic of the ion and (ii) the difference between the energy of ion in its equilibrium geometry and in the neutral relaxed geometry.

2.1.3.2 Normal Mode Analysis

The normal mode analysis provides an approach to obtain the total relaxation energy from the contributions of each vibrational mode: [9]

$$\lambda_i = \frac{k_i}{2}\Delta Q_i^2 = \hbar\omega_i S_i \tag{2.13}$$

$$\lambda = \sum_i \lambda_i \tag{2.14}$$

Here, i runs over all the vibrational normal modes (NM), k_i and ω_i are the corresponding force constant and frequency, ΔQ_i represents the displacement between the equilibrium geometries of the neutral and charged molecules, and S_i denotes the Huang-Rhys factor measuring the electron–phonon coupling strength for the ith normal mode.

2.1.4 Mobility Evaluation

With the knowledge of the charge transfer rates between neighboring molecules, one can evaluate the charge mobility simply from the information of one single hopping step, or more accurately through random walk simulation for the charge diffusion trajectories.

2.1.4.1 Single-Step Approximation with Electric Field

In principle, the charge carrier mobility (μ) can be obtained from its definition as the ratio between the charge drift velocity (v) and the driving electric field (F): [7]

$$\mu = \frac{v}{F} \tag{2.15}$$

Assuming that the charge transport is a Boltzmann hopping process and one pathway with only one single hopping step can characterize the whole diffusion properties, v can be approximately calculated from the nearest inter-site distance (a) and the hopping time (τ), which is actually the reciprocal of the charge transfer rate (k):

$$v \simeq a/\tau = ak \tag{2.16}$$

Note that within this approach, an additional contribution from the electric field should be added to the free energy part of Eq. 2.1:

$$k = \frac{V^2}{\hbar}\sqrt{\frac{\pi}{\lambda k_B T}}\exp\left\{-\frac{(\Delta G^0 + eFa + \lambda)^2}{4\lambda k_B T}\right\} \tag{2.17}$$

2.1.4.2 Single-Step Approximation with Einstein Relation

The Einstein relation sets up the relationship between the mobility and the diffusion constant (D): [10, 11]

$$\mu = \frac{e}{k_B T} D \tag{2.18}$$

$$D = \frac{1}{2d}\frac{\langle l(t)^2 \rangle}{t} \tag{2.19}$$

Here, $l(t)$ is the distance between the charge position at time t and its starting point at time zero. Earlier treatment for the hopping mobility was obtained as [12]

$$D = a^2 k \tag{2.20}$$

where a is the intermolecular spacing and k is the charge transfer rate, namely, the inverse of hopping time. Such an approach requires only a molecular dimer to estimate the bulk mobility. Later, this model has been extended to average over all the neighbors: [13]

$$D \approx \frac{1}{2d}\sum_i P_i \cdot r_i^2 k_i \tag{2.21}$$

Here, d is the space dimension, namely, $d = 1, 2, 3$ for 1D, 2D and 3D systems, respectively. The index i covers all the hopping pathways out of a chosen reference molecule with r_i being the corresponding hopping distance, which is usually expressed as the intermolecular center to center distance, and k_i being the charge transfer rate. P_i is the relative probability to choose the ith pathway:

$$P_i = k_i \Big/ \sum\nolimits_j k_j \tag{2.22}$$

2.1.4.3 Random Walk Monte Carlo Simulation

The mobility is a bulk parameter, therefore it should be strongly related to the long range molecular packing, which may provide entirely different transport networks and give different results compared with the simple estimations in previous two sections considering only one single hopping step, especially for inhomogeneous system [14]. To solve this problem, we should model the Brownian motion of charge transport explicitly. Thereby, we propose a random walk scheme to simulate the charge diffusion using the Monte Carlo technique. First, we take the experimental measured crystal structure exactly for simulation and choose an arbitrary molecule within the bulk as the starting point for the charge. The charge is only allowed to hop to its nearest neighbor molecules. To decide which the next site is for the charge to land in, a random number r uniformly distributed between 0 and 1 is generated. Then the charge is allowed to go along the ith specified direction when $\sum_{j=1}^{i-1} P_j < r \le \sum_{j=1}^{i} P_j$, where P_j is the relative hopping probability given by Eq. 2.22. The hopping distance is taken to be the intermolecular center distance of the corresponding pathway and the hopping time is set to $1/\sum_j k_j$. Such Monte Carlo simulation keeps going until the diffusion distance exceeds at least 10^2–10^3 times the lattice constant. Following the same process, thousands of simulations should be performed to get a linear relationship between the mean-square displacement and the simulation time to get the diffusion

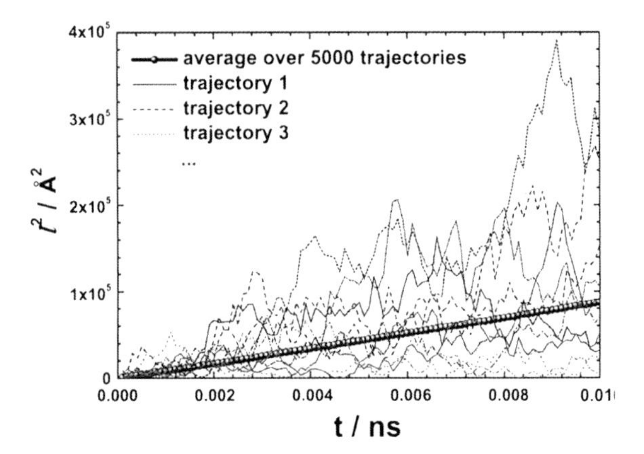

Fig. 2.2 A typical time-dependent squared displacement from random walk simulations. Reproduced from Ref. [42] by permission of the PCCP Owner Societies

coefficient according to Eq. 2.19 (see Fig. 2.2). The mobility is finally calculated through the Einstein relation in Eq. 2.18. It has been shown that two thousand simulations are enough to get converged results of mobility [9]. Note that organic crystals generally have layer-by-layer ordered structures with weak electronic couplings between layers and much larger electronic coupling within each layer, thus the isotropic diffusion assumed in Sect. 2.1.4.2 is generally not valid, but the Monte Carlo procedure can be used to simulate the anisotropic mobilities when the charge trajectories are projected to a specified lattice direction.

2.2 Application I: Siloles

The silicon containing cyclic π-conjugated system silole (silacyclopentadiene) is a promising emissive material because of its notable aggregation-enhanced emission [15, 16] and high PL efficiency [17, 18] in thin solid films. Siloles are also believed to be excellent electron transport materials because the presence of the silicon atom lowers the LUMO energy level relative to that of pure hydrocarbon molecules and this facilitates electron injection. The lowering of the LUMO has been ascribed to the interaction between the σ^* orbital of two exocyclic σ-bonds on the silole ring and the π^* orbital of the butadiene moiety [19, 20]. Here, we apply the methods described in Sect. 2.1 to investigate the charge transport properties in two silole-based compounds, i.e., 1,1,2,3,4,5-hexaphenylsilole (HPS) and 1-methyl-1,2,3,4,5-pentaphenyl-silole (MPPS), as shown in Fig. 2.3.

2.2.1 Computational Details

We adopt the single crystal structures of HPS [21] and MPPS [22] from experiment and generate a variety of possible intermolecular hopping pathways. The unit cells of both HPS and MPPS contain two inequivalent molecules, namely, type A and B. As a result, there are two kinds of pathways, i.e., A–A, and A–B, if we take

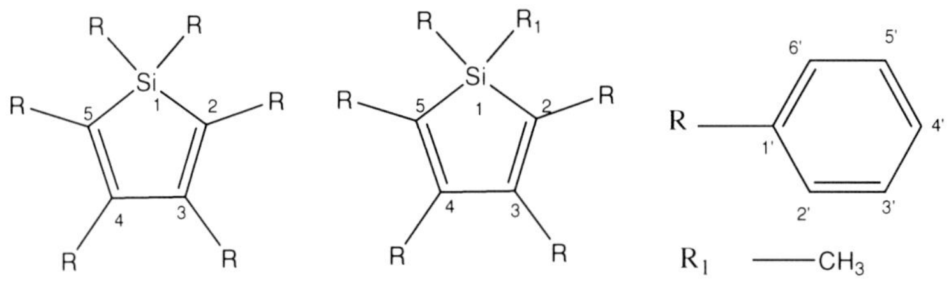

HPS MPPS

Fig. 2.3 Chemical structure of HPS (*left*) and MPPS (*right*). Reprinted with permission from Ref. [7]. Copyright 2006 American Chemical Society

molecule A as the reference. The first occurs only between cells, while the latter can occur both within and between cells. Using the primitive cell (0, 0, 0) as a reference, all the pathways can be defined with their cell indexes (h, k, l) and molecule types. The intermolecular transfer integrals are calculated based on the direct method according to Eq. 2.2. The internal reorganization energies are calculated with the diabatic energy surface approach as shown in Sect. 2.1.3.1. All these calculations are performed with the Gaussian 03 package [23]. The external reorganization energies are set to be 0.2–0.6 eV. The mobilities are calculated with the simple model introduced in Sect. 2.1.4.1.

2.2.2 Results and Discussion

2.2.2.1 Transfer Integral

The calculated transfer integrals for HPS and MPPS crystals are listed in Tables 2.1 and 2.2, respectively. For HPS, the largest transfer integral for electron and hole are found to be 17.69 meV and 14.10 meV, respectively which are quite close with each other. For MPPS, the situation is quite similar (33.47 vs. 43.97 meV), but the transfer integrals are generally much larger than HPS. This can be easily understood by their difference in intermolecular distances (see Table 2.3). For the most important pathways, the inter-carbon distances in MPPS are around 6.71–6.86 Å, which are much smaller than those of HPS within the range of 9.16–11.74 Å. This is due to the fact that the phenyl group presents larger hindrance than the methyl group and thus MPPS has greater π–π overlap. Besides, we can find that there exist channels where electron transfer is much larger than that for the hole, e.g., channels I, VI, and X for HPS and channels I, VIII, and IX for MPPS, while there are just few channels that hole transfer integral is larger than electron. This can be explained by the charge distributions of the frontier orbitals. Taking HPS as an example, the LUMO wave function is mainly localized on the silole ring, especially on the 2-, 1-, and 5-positions (see Fig. 2.4), and there is also considerable electron density on the 1′-position of the silicon exocyclic aryl ring, which is due to the interaction between the σ^* orbitals of the two exocyclic σ

Table 2.1 Charge transfer integrals (t_h for hole and t_e for electron) between molecule A in HPS (0, 0, 0) unit cell and the other most possible adjacent molecules

Pathway	A–A			A–B							
Channel	I	II	III	IV	V	VI	VII	VIII	IX	X	XI
Partner	(± 1, 0, 0)	(0, ±1, 0)	(0, 0, ±1)	(0, 0, 0)	(−1, 0, 0)	(0, −1, 0)	(0, 0, −1)	(−1, −1, 0)	(−1, 0, −1)	(0, −1, −1)	(−1, −1, −1)
Si–Si distance (Å)	9.53	10.04	16.31	7.20	8.70	8.78	9.99	12.40	13.18	13.85	18.03
t_h (meV)	0.79	0.46	0.03	17.69	0.08	2.39	2.26	0.05	4.87	2.39	0.03
t_e (meV)	6.61	1.17	0.03	12.82	1.25	14.10	1.85	0.14	4.41	12.05	0.19

Table 2.2 Charge transfer integrals (t_h for hole and t_e for electron) between molecule A in MPPS (0, 0, 0) unit cell and the other most possible adjacent molecules

Pathway	I (A–A)			II (A–B)							
Channel	I	II	III	IV	V	VI	VII	VIII	IX	X	XI
Partner	(±1, 0, 0)	(0, ±1, 0)	(0, 0, ±1)	(0, 0, 0)	(−1, 0, 0)	(0, −1, 0)	(0, 0, −1)	(−1, −1, 0)	(−1, 0, −1)	(0, −1, −1)	(−1, −1, −1)
Si–Si distance (Å)	10.19	11.27	11.98	13.72	7.48	15.33	10.66	10.42	6.10	12.88	9.76
t_h (meV)	7.70	1.09	4.54	0.03	33.47	5.82	0.54	4.38	2.10	0.98	3.10
t_e (meV)	15.08	0.54	1.39	0.08	43.97	4.08	0.49	7.95	18.04	1.74	1.80

Table 2.3 Interatomic distances (in Å) in the molecular dimers for HPS and MPPS

Channel	Si1–Si1	C2–C2	C5–C5	C3–C3	C4–C4
HPS-IV	7.61	9.63	9.63	11.74	11.74
HPS-VI	8.78	9.16	9.16	10.26	10.26
MPPS-V	7.48	6.71	6.71	6.86	6.86
MPPS-IX	6.10	8.48	8.48	10.82	10.82

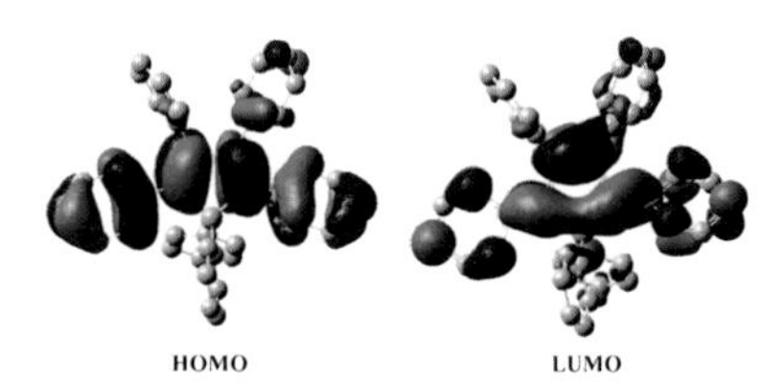

Fig. 2.4 HOMO and LUMO of HPS molecule. Reprinted with permission from Ref. [7]. Copyright 2006 American Chemical Society

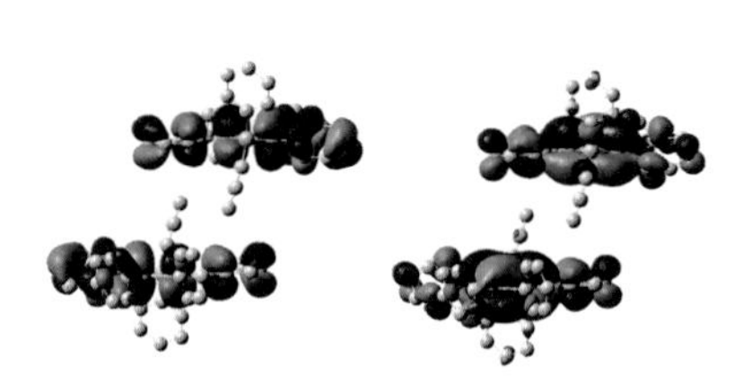

Fig. 2.5 HPS dimer structures, including frontier molecular orbitals of (0, 0, 0) A (*down*) and (0, −1, 0) B (*up*) molecules. The *left panel* shows the HOMOs and the *right panel* shows in LUMOs. Reprinted with permission from Ref. [7]. Copyright 2006 American Chemical Society

Table 2.4 Calculated internal reorganization energies (in eV) for MPPS and HPS

		Neutral to ion	Anion to ion	Total
MPPS	Electron	0.18	0.16	0.34
	Hole	0.14	0.13	0.27
HPPS	Electron	–	–	0.32
	Hole	–	–	0.27

orbital on the ring silicon and the π^* orbital of the butadiene moiety [24]. On the other hand, the HOMO (hole) is primarily located on the C2=C3 and C4=C5 double bonds in the silole ring as well as on the 2- and 5-positions in the aryl rings. The different contribution of the LUMO and HOMO orbitals may induce different relative strength of overlap between molecules for different pathways. For example, in the configuration of channel VI (see Fig. 2.5), the distances between silole rings at the 1-, 2-, and 5-positions of the $(0, 0, 0)$ A molecule and the $(0, -1, 0)$ B molecule are smaller than those at the 3- and 4-positions, which eventually make the intermolecular overlap for LUMOs larger than that for HOMOs.

2.2.2.2 Reorganization Energy

The calculated internal reorganization energies are given in Table 2.4. We can find that the total reorganization energies for MPS and HPS are similar, and the

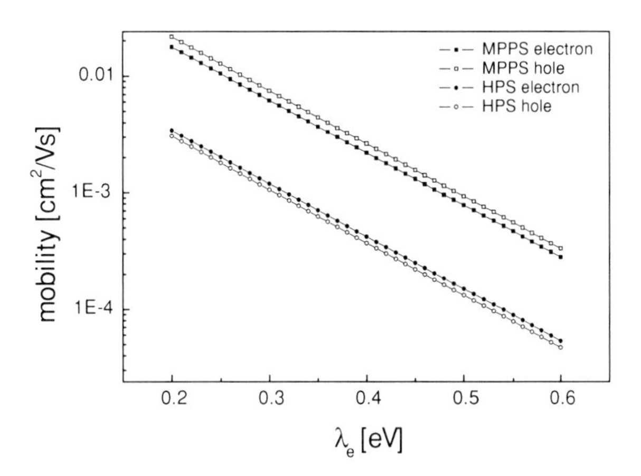

Fig. 2.6 Theoretical estimation of the room-temperature carrier mobilities for MPPS and HPS for the most favored transport channel, with an applied electric field of $F = 10^6$ V/cm. Reprinted with permission from Ref. [7]. Copyright 2006 American Chemical Society

reorganization energy for the hole is slightly smaller than that for the electron in both MPPS and HPS.

2.2.2.3 Room-Temperature Mobility

The largest transfer rates (channel IV for both HPS and MPPS) are used to calculate the mobilities. In Fig. 2.6, we show the room-temperature mobility as a function of the external reorganization energy for an electric field of $F = 10^6$ V/cm, which is a typical value for organic light emitting diode devices. An interesting discovery is that, although in most cases, electron mobility is observed to be much smaller than hole mobility in organic materials, the calculated mobilities of electron and hole are about the same in HPS and MPPS. This can be explained by the fact that the reorganization energy of hole is smaller than electron and more transport channels exist for electron than hole. Such balanced electron and hole mobilities in siloles should be one of the reasons for the high electroluminescence efficiency. Besides, the mobility of MPPS is generally one magnitude larger than that of HPS because the transfer integral of MPPS is larger than HPS and their reorganization energies are quite close.

2.3 Application II: Triphenylamine Dimers

Triphenylamine (TPA) derivatives have been widely employed as hole transport materials in molecular electronics applications [25, 26]. However, their performances have been found to be unsatisfactory due to their amorphous nature in the solid state [27–29]. In order to improve the mobility of TPA, a design strategy of making dimers of TPA, either in the form of a macrocycle or a linear chain (see Fig. 2.7), has been performed [30, 31]. This part tends to understand the origin of their mobility-structure relationship from bottom up.

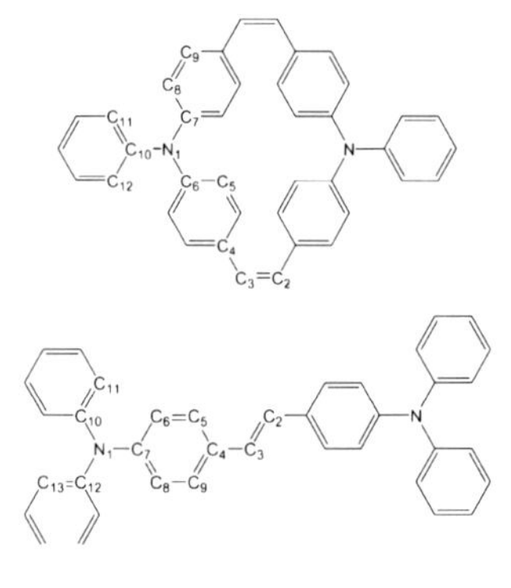

Fig. 2.7 Chemical structure of the cyclic triphenylamine (denoted as compound 1, *upper panel*), and a linear analog (denoted as compound 2, *lower panel*). Reproduced from Ref. [31] by permission of the AAS

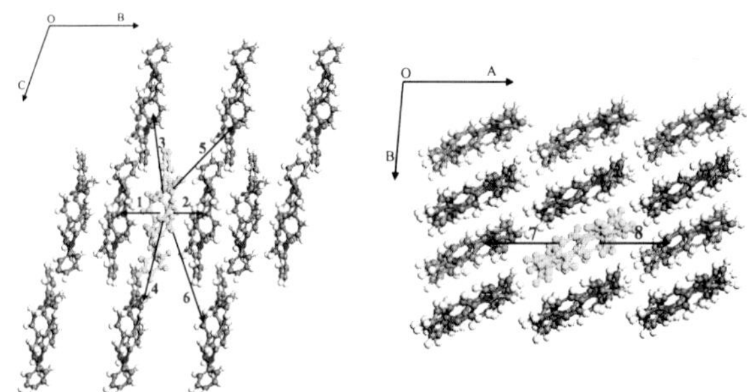

Fig. 2.8 Hopping pathways for compound 1. Reprinted with permission from Ref. [30]. Copyright 2006 American Chemical Society

2.3.1 Computational Details

The crystal structures are taken from experimental X-ray analyzed crystallographic data [30]. The transfer integrals are evaluated within direct scheme of Kohn-Sham–Fock operator as described in Sect. 2.1.2.1. The reorganization energy is obtained with the diabatic potential surfaces. And the mobility is calculated from single-step approach with Einstein relation introduced in Sect. 2.1.4.2. All quantum chemistry calculations are performed within Gaussian03 package [23].

2.3.2 Results and Discussion

2.3.2.1 Transfer Integral

The chosen pathways are shown in Figs. 2.8 and 2.9 for compound 1 and 2, respectively. And the corresponding transfer integrals are given in Tables 2.5 and 2.6. Since there are two molecules in the unit cell of compound 2, two sets of data are given corresponding to choosing either of them as the reference molecule. It is found that the calculated two sets of transfer integrals do not differ much, e.g., the largest transfer integral is 8.65×10^{-3} and 6.26×10^{-3} eV, respectively. Besides, the magnitude of the transfer integrals for both compounds is also quite close.

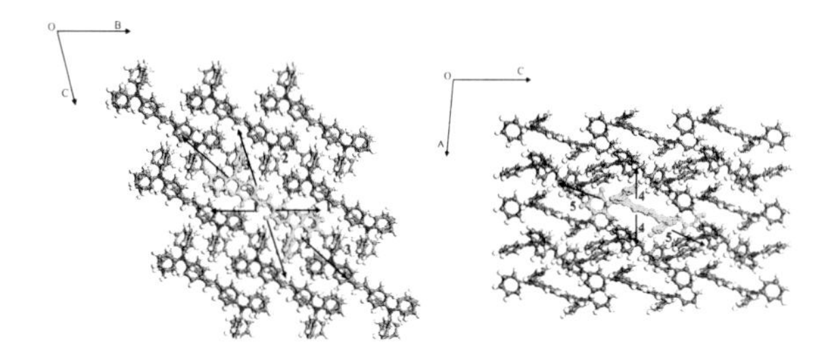

Fig. 2.9 Hopping pathways for compound 2. Reprinted with permission from Ref. [30]. Copyright 2006 American Chemical Society

Table 2.5 Calculated transfer integrals for eight pathways in compound 1

Pathway	Distance (Å)	V (eV)
1	5.133	4.86E-3
2	5.328	8.65E-3
3	13.502	2.50E-3
4	13.974	7.43E-3
5	15.927	1.36E-7
6	15.406	1.21E-8
7	10.642	4.09E-3
8	10.642	4.09E-3

Table 2.6 Calculated two sets of transfer integrals for compound 2

Set	Pathway	Distance (Å)	V (eV)
1	1	11.334	4.35E-3
	2	12.491	9.97E-5
	3	19.102	2.19E-4
	4	5.312	5.64E-3
	5	13.575	3.03E-3
2	1	11.334	6.26E-3
	2	12.491	1.68E-3
	3	19.102	3.08E-3
	4	5.312	5.67E-3
	5	13.575	3.03E-7

2.3.2.2 Reorganization Energy

For compound 1, the reorganization energy is calculated to be 0.173 eV, which is much smaller than that of compound 2, 0.317 eV. This difference in internal reorganization energy can be explained through the structure difference between optimized neutral molecule and cation (see Table 2.7). From the difference in bond lengths, bond angles, and the torsions, we can find that the change in geometry from neutral molecule to cation is smaller in compound 1 than compound 2, since

Table 2.7 The optimized geometrical parameters for the neutral and cation structures of compounds 1 and 2, including bond lengths, bond angles, and dihedral angles values

		Neutral	Cation	Δ
Compound 1	C_2C_3	1.35 Å	1.36 Å	0.01 Å
	C_3C_4	1.48 Å	1.46 Å	−0.02 Å
	N_1C_6, N_1C_7	1.42 Å	1.41 Å	−0.01 Å
	C_7C_8	1.40 Å	1.41 Å	0.01 Å
	C_8C_9	1.40 Å	1.39 Å	−0.01 Å
	$C_{10}C_{11}$, $C_{10}C_{12}$	1.41 Å	1.40 Å	−0.01 Å
	$C_{10}N_1C_7$, $C_{10}N_1C_6$	121.3°	121.1°	−0.2°
	$C_6N_1C_7$	117.4°	117.9°	0.5°
	$C_{11}C_{10}N_1C_7$	35.1°	40.7°	5.6°
	$C_5C_6N_1C_7$	44.0°	40.2°	−3.8°
	$C_8C_7N_1C_{10}$	46.1°	41.3°	−4.8°
Compound 2	C_2C_3	1.35 Å	1.37 Å	0.02 Å
	C_3C_4	1.46 Å	1.43 Å	−0.03 Å
	C_4C_5, C_4C_9	1.41 Å	1.42 Å	0.01 Å
	C_5C_6, C_8C_9	1.39 Å	1.38 Å	−0.01 Å
	C_6C_7	1.41 Å	1.42 Å	0.01 Å
	C_7C_8	1.40 Å	1.42 Å	0.02 Å
	N_1C_7	1.42 Å	1.38 Å	−0.04 Å
	N_1C_{10}, N_1C_{12}	1.42 Å	1.43 Å	0.01 Å
	$C_4N_1C_3$	120.1°	121.3°	0.2°
	$C_4N_1C_2$	119.7°	117.5°	−2.2°
	$C_2N_1C_3$	120.2°	121.2°	1.0°
	$C_{11}C_{10}N_1C_7$	42.2°	51.2°	10.0°
	$C_{13}C_{12}N_1C_{10}$	42.5°	51.2°	8.7°
	$C_6C_7N_1C_{10}$	39.6°	24.0°	−15.6°

the closed ring structure restricts the rotation of the phenyl groups. As a result, lower reorganization energy is obtained.

2.3.2.3 Room-Temperature Mobility

The average room-temperature hole mobilities are calculated to be 2.7×10^{-2} cm^2/Vs for compound 1 and 1.9×10^{-3} cm^2/Vs for compound 2. These values compare quite well with the experimental results of $0.5–1.5 \times 10^{-2}$ cm^2/Vs for 1 and 2×10^{-4} cm^2/Vs for 2, respectively, except that the theoretical value for 2 is about one order of magnitude larger than the experimental value. It seems that the experimental mobility for 2 can still be improved by subsequent processing and optimization of the material. Since the two compounds are similar in transfer integrals, the difference in mobility can be attributed to their difference in reorganization energy. Therefore, material with macro-cyclic structure is more favorable for charge transport than linear structure due to smaller reorganization energy during charge transport processes.

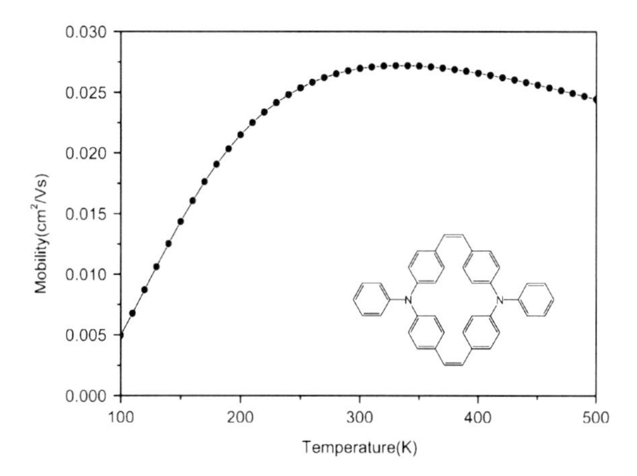

Fig. 2.10 The predicted temperature dependence of the drift mobility for compound 1. Reproduced from Ref. [31] by permission of the AAS

2.3.2.4 Temperature Dependence of Mobility

The temperature dependence of mobility is an important topic to understand the charge transport mechanism in organic materials. Here, we show how Marcus theory describes this temperature dependence, using compound 1 and 2 as examples. We assume that both the transfer integrals and the reorganization energy do not change with temperature, but we note that the impact of such change will be detailedly investigated for another organic material in Chap. 3. Combining Eqs. 2.1, 2.18, and 2.20, it is easy to see that the mobility should have a temperature dependence of $T^{-3/2}\exp(-\lambda/4k_BT)$. The exponential law, $\exp(-\lambda/4k_BT)$, dominates at low temperatures and the power law, $T^{-3/2}$, dominates at high temperature after the barrier is fully overcome, namely, the mobility increases with the increase of temperature at low T and decreases with temperature at high T (see Fig. 2.10). The maximum of mobility is directly related to the barrier height of $\lambda/4$. It should be noted that Marcus theory can only be applied at high temperature since the quantum tunneling effect of the nuclear motions, which is very important at low temperatures, has been neglected. The improvement of adding quantum effects inside the hopping picture will be discussed in Sect. 2.4. In Chaps. 3 and 4, the temperature dependence of mobility will also be discussed in-depth using other theoretical models.

2.4 Application III: Oligothiophenes

Oligothiophene (nT) is one of the earliest organic materials for organic thin film field-effect transistors (FETs) [32, 33]. Thiophene-based materials exhibit a variety of intra- and intermolecular interactions originating from the high polarizability of sulfur electrons in the thiophene rings [34]. Therefore, thiophene oligomers can be regarded as a versatile building block for organized structures. The crystals of oligothiophenes exhibit a herringbone structure with different number of molecules in the unit cell (Z), depending on the sublimation temperatures (see

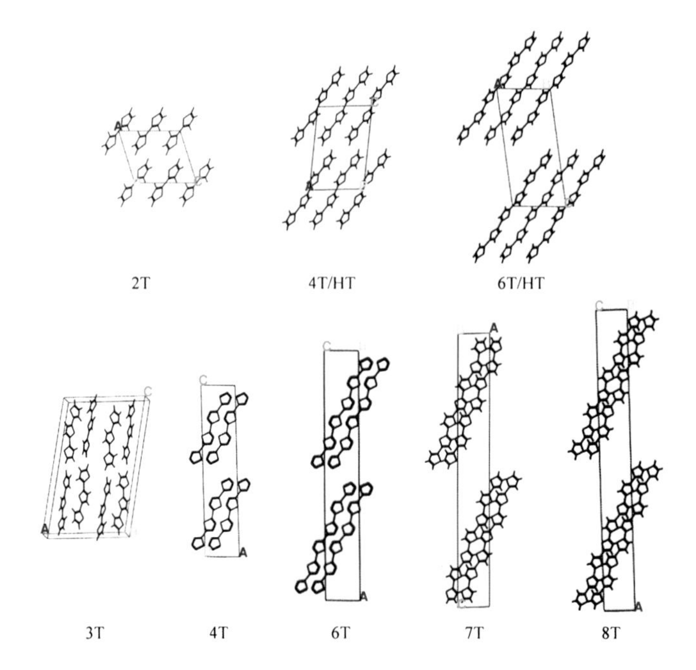

Fig. 2.11 Crystal structures of α-*nTs*. Reprinted with permission from Ref. [9]. Copyright 2008 American Chemical Society

Fig. 2.11). For the high-temperature (HT) phase, $Z = 2$ for 2, 4, and 6T, while for low temperature (LT) phase, $Z = 4$ for 4, 6, 7, and 8T. In addition, for 3T, $Z = 8$. Using *n*T as a series of model systems, hereby, we discuss the influence of crystal structure and molecular size on mobility.

2.4.1 Computational Details

The transfer integrals are calculated within the direct scheme and the reorganization energies are obtained through the NM analysis. The Huang–Rhys factors are evaluated through the DUSHIN program [35]. After obtaining the Marcus transfer rates, the mobility is calculated with a random walk simulation, which is introduced in Sect. 2.1.4.3.

2.4.2 Results and Discussion

2.4.2.1 Transfer Integral

The oligothiophenes with the same *Z* has very similar crystal structures. Thus we only show the chosen hopping pathways for 4T/HT ($Z = 2$), 4T/LT ($Z = 4$), and

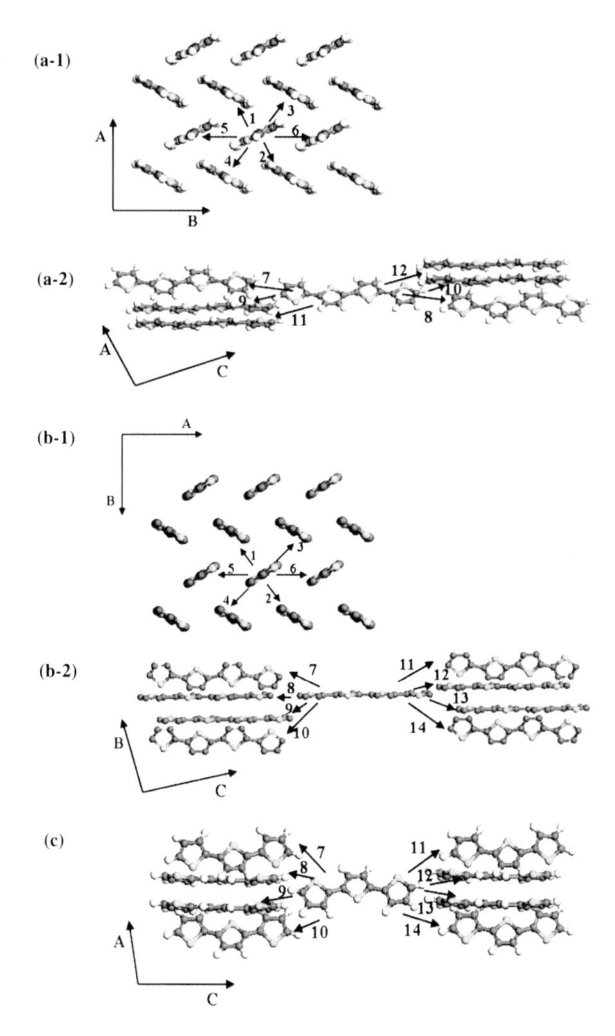

Fig. 2.12 Charge hopping pathways schemes for 4T/HT (**a**), 4T/LT (**b**), and 3T (**c**). **a-1** and **b-1** display the hopping routes in the same molecular layer, and (**a-2**) and (**b-2**) are between different molecular layers. Reprinted with permission from Ref. [9]. Copyright 2008 American Chemical Society

3T ($Z = 8$) in Fig. 2.12. The calculated transfer integrals are given in Tables 2.8 and 2.9. There, we can find that the transfer integrals are not influenced by the oligomer length within the same group of packing. Besides, the largest transfer integral for $Z = 2$ crystals is about 34–40 meV, which is about twice as much as that for the $Z = 4$ phase, which is only about 18 meV. To understand this phenomenon, we illustrate explicitly the packing structures for the dimer along the principal pathway for the HT and LT phases of 4T and 6T, as well as their HOMO coefficients (see Fig. 2.13). As is known, the transfer integral is increased if both are bonding or antibonding interactions between the π-atomic orbitals and decreased when there occurs a cancelation between bonding and antibonding overlap. It is noted that for the LT phase, there exists a displacement of about half a thiophene ring width, while for the HT phase, the displacement is about one

Table 2.8 Calculated transfer integrals (V, in meV) and intermolecular distances (d, in Å) for all pathways for 4T/LT, 4T/HT, and 6T/HT at the DFT/PW91PW91/6-31G* level

Pathway	2T		4T/HT		6T/HT	
	V	d	V	d	V	d
1	34	5.34	40	5.31	36	5.38
5	6	5.90	4	5.75	3	5.68
7	1	10.00	0.7	17.82	0.4	25.68
9	12	8.31	2	15.81	0.7	23.38

Table 2.9 Calculated transfer integrals (V, in meV) and intermolecular distances (d, in Å) for all pathways of other than $Z = 2$ crystal phases for 3, 4, 6, 7, and 8T at the DFT/PW91PW9 1/6-31G* level

Pathway	3T		4T		6T		7T		8T	
	V	d	V	d	V	d	V	d	V	d
1	7	4.67	12	4.93	18	4.98	18	4.81	17	4.92
2	13	5.06	17	5.01	16	4.92	14	4.97	17	4.96
3	13	4.72	17	5.01	16	4.92	17	4.82	17	4.96
4	13	5.06	12	4.93	18	4.98	18	4.97	17	4.92
5	0.4	5.64	18	6.08	10	6.03	13	5.95	9	6.00
6	0.4	5.64	18	6.08	10	6.03	13	5.95	9	6.00
7	2	12.16	5	16.49	2	24.18	0.4	28.82	1	31.88
8	4	13.25	3	17.63	2	25.34	0.4	29.67	0	30.73
9	4	13.20	4	15.55	3	23.16	1	27.68	0.5	32.95
10	19	12.29	0.0	16.96	0.0	24.51	0.3	28.84	0	32.13
11	0	14.87	5	16.49	2	24.18	1	29.17	1	31.88
12	4	13.20	0.2	15.41	0.1	23.04	2	27.05	1	33.05
13	4	13.25	2	17.46	1	25.22	0	28.37	1	30.82
14	0.3	14.53	0	16.96	0	24.51	2	28.07	0	32.13

thiophene ring. Combined with the HOMO orbital charge distributions, one can easily rationalize that the molecular packing in the HT phase favors hole transfer better than the LT phase. And their transfer integral can differ about 2–3 times even though the intermolecular distances are almost the same.

2.4.2.2 Reorganization Energy

From Table 2.10, it is easy to find that the calculated reorganization energies from the adiabatic potential surface method are very similar with those from the normal mode analysis. This indicates that the molecular reorganizing process can be well-described by the harmonic oscillator model as assumed in Marcus theory. It is also seen that the reorganization energy decreases as the chain is elongated. To understand the reason, we display the partition of the relaxation energies of 4T and 8T into the contributions of each normal mode, which is shown in Fig. 2.14. We can see that the contributions from the high-frequency parts (1,200–1,600 cm^{-1})

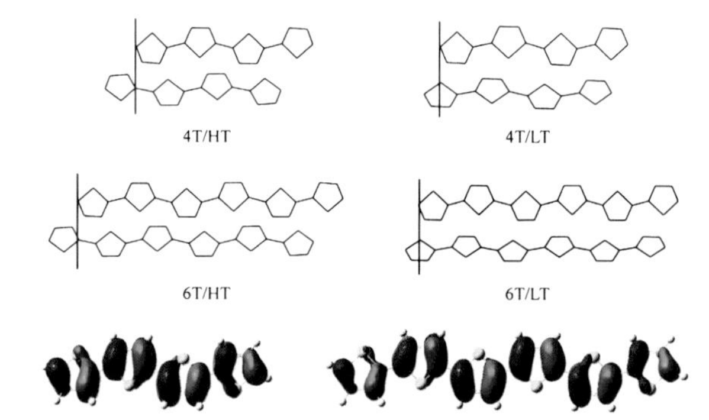

Fig. 2.13 Intermolecular displacements taken from crystal packing along the long axis of thiophene for the dominant pathway. HOMOs of 4T and 6T also are shown. Reprinted with permission from Ref. [9]. Copyright 2008 American Chemical Society

Table 2.10 Calculated reorganization energies by adiabatic potential surfaces of the neutral and cation species and by NM analysis at the DFT-B3LYP/6-31G* level

Molecules	Reorganization energy	
	Adiabatic potential	Normal mode
2T	361	364
3T	316	323
4T	286	288
5T	265	274
6T	244	255
7T	224	238
8T	203	212

decrease remarkably when going from 4T to 8T, which means that the C–C single and double stretching modes are influenced by the conjugation length [36].

2.4.2.3 Mobility

The calculated mobilities are shown in Fig. 2.15. Within the group of *nT* crystals with the same *Z*, the mobility increases with the number of thiophene rings, n, since they have close transfer integrals, but the reorganization energy decreases with the chain length. Besides, the HT phase ($Z = 2$) leads to a larger drift mobility than the LT phase ($Z = 4$). This is because the former has larger transfer integral than the latter, while the reorganization energy is the same.

2.5 Incorporate Nuclear Tunneling Effect in the Hopping Picture

The Marcus charge transfer rate works for extreme high-temperature regimes. At room temperature, $k_B T$ is about 200 cm^{-1}, which is much smaller than some of the high-frequency vibrations, e.g., the single and double bond stretching modes with frequency 1,200–1,600 cm^{-1} (as shown in Fig. 2.14). Therefore the nuclear

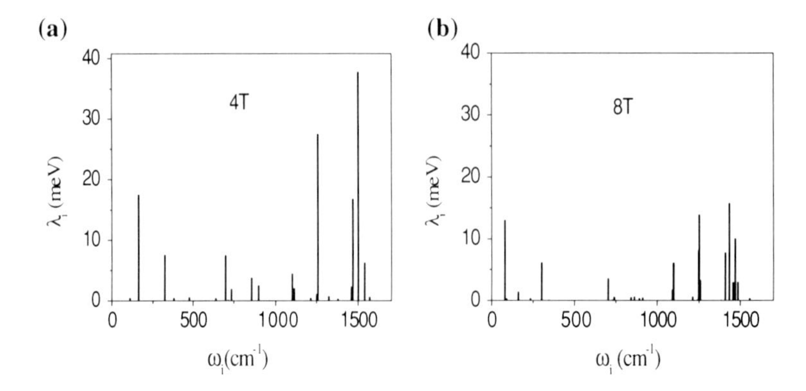

Fig. 2.14 Contribution of the vibrational modes to the cationic relaxation energy for 4T (**a**) and 8T (**b**). Reprinted with permission from Ref. [9]. Copyright 2008 American Chemical Society

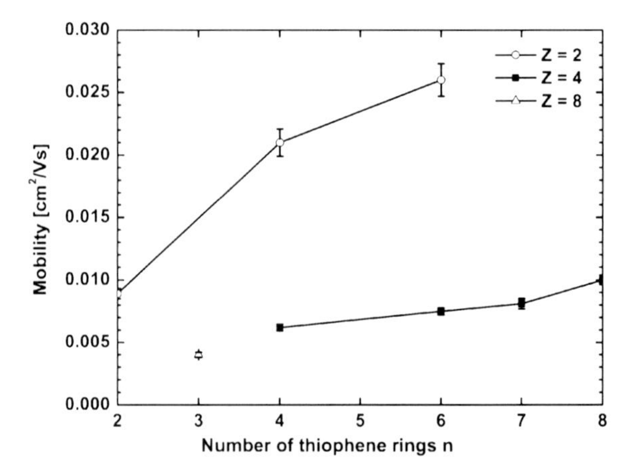

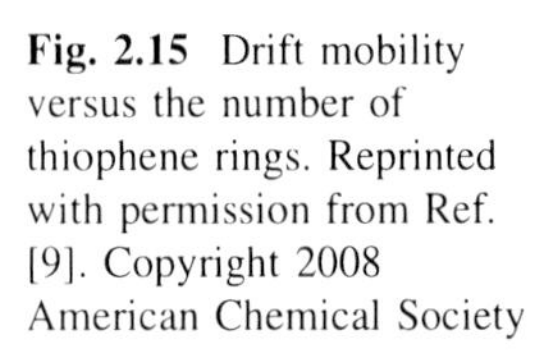
Fig. 2.15 Drift mobility versus the number of thiophene rings. Reprinted with permission from Ref. [9]. Copyright 2008 American Chemical Society

tunneling effect of these modes needs to be taken into account in calculating the charge transfer rate. In following, we will use the Fermi's Golden rule to obtain the charge transfer rate, and show the role of nuclear tunneling on charge carrier mobility.

2.5.1 Fermi's Gold Rule

The charge transfer rate formula considering the quantum effects was derived by Jortner [37] and Lin et al. [38] It starts from the general Fermi's Golden rule for the transfer rate from the initial state, $|i\rangle$, to the final state, $|f\rangle$:

$$k = \frac{2\pi}{\hbar^2} V_{fi}^2 \delta(\hbar\omega_{fi}) \tag{2.23}$$

Here, V_{fi} and $\hbar\omega_{fi}$ are the coupling and energy difference between the final and initial state. If we write the contributions of the electronic state and the nuclear vibrational state separately, we get

$$k = \frac{2\pi}{\hbar^2} V^2 \sum_{\upsilon,\upsilon'} P_{i\upsilon} \left| \langle \Theta_{f\upsilon'} \mid \Theta_{i\upsilon} \rangle \right|^2 \delta\left(\hbar\omega_{f\upsilon',i\upsilon}\right) \tag{2.24}$$

Now V is the transfer integral described previously in Sect. 2.1, i.e., the coupling between the final and initial electronic states, υ (υ') is the quanta of the nuclear vibration in the initial (final) state, $P_{i\upsilon}$ is the distribution function of υth vibrational quanta in the initial state, $\Theta_{i\upsilon(f\upsilon')}$ is the wave function of the initial (final) nuclear vibration, and $\hbar\omega_{f\upsilon',i\upsilon}$ is the energy different between the final and initial vibronic states. If the nuclear vibration consists of a collection of independent harmonic oscillators, we have

$$\Theta_{i\upsilon} = \prod_j \chi_{i\upsilon_j}(Q_j) \tag{2.25}$$

$$\Theta_{f\upsilon'} = \prod_j \chi_{f\upsilon'_j}\left(Q'_j\right) \tag{2.26}$$

$$P_{i\upsilon} = \prod_j P_{i\upsilon_j} \tag{2.27}$$

$$\omega_{f\upsilon',i\upsilon} = \omega_{fi} + \sum_{\upsilon_j} \sum_{\upsilon'_j} \left[\left(\upsilon'_j + \frac{1}{2} \right) \omega'_j - \left(\upsilon_j + \frac{1}{2} \right) \omega_j \right] \tag{2.28}$$

where

$$\chi_{\upsilon_j}(Q_j) = \left(\beta_j / \sqrt{\pi} 2^{\upsilon_j} \upsilon_j! \right)^{1/2} H_{\upsilon_j}(\beta_j Q_j) \exp\left(-\beta_j^2 Q_j^2 / 2 \right) \tag{2.29}$$

$$\begin{aligned} P_{i\upsilon} &= \left[\sum_{\upsilon} \exp\left(\frac{-E_{i\upsilon}}{k_B T} \right) \right]^{-1} \exp\left(\frac{-E_{i\upsilon}}{k_B T} \right) \\ &= \prod_j^N 2 \sinh \frac{\hbar\omega_j}{2k_B T} \exp\left(-\hbar\omega_j \left(\upsilon_j + \frac{1}{2} \right) / k_B T \right) \end{aligned} \tag{2.30}$$

with $\beta_j = (\omega_j/\hbar)^{1/2}$ and $H_{\upsilon j}$ are the Hermite polynomials. Expressing the δ function as a Fourier integral in time, Eq. 2.24 becomes

$$k = \frac{V^2}{\hbar^2} \int_{-\infty}^{\infty} dt e^{it\omega_{fi}} \prod_j G_j(t) \tag{2.31}$$

$$G_j(t) = \sum_{\upsilon_j} \sum_{\upsilon'_j} P_{i\upsilon_j} \left| \left\langle \chi_{f\upsilon'_j} \mid \chi_{i\upsilon_j} \right\rangle \right|^2 \exp\left(it \left\{ \left(\upsilon'_j + \frac{1}{2} \right) \omega'_j - \left(\upsilon_j + \frac{1}{2} \right) \omega_j \right\} \right) \quad (2.32)$$

After employing the Slater sum and the displaced harmonic oscillator approximation [39], Eq. 2.32 can be evaluated as:

$$G_j(t) = \exp\left[-S_j \left\{ (2n_j + 1) - n_j e^{-it\omega_j} \omega_j - (n_j + 1) e^{it\omega_j} \right\} \right]. \quad (2.33)$$

where $n_j = 1/(\exp(\hbar\omega_j/k_BT) - 1)$ and $S_j = \lambda_j/\hbar\omega_j$ are the population and the Huang–Rhys factor of the jth normal mode. Substituting Eq. 2.33 into Eq. 2.31 yields the quantum charge transfer rate expression:

$$k = \frac{V^2}{\hbar^2} \int_{-\infty}^{\infty} dt \exp\left\{ it\omega_{fi} - \sum_j S_j \left[(2n_j + 1) - n_j e^{-it\omega_j} - (n_j + 1) e^{it\omega_j} \right] \right\} \quad (2.34)$$

Finally, we take the real part of the integration of Eq. 2.34 and get:

$$k = \frac{V^2}{\hbar^2} \int_0^{\infty} dt \exp\left\{ -\sum_j S_j (2n_j + 1)(1 - \cos \omega_j t) \right\} \cos\left(\sum_j S_j \sin \omega_j t \right) \quad (2.35)$$

2.5.2 *Short-Time Integration in Fermi's Golden Rule*

We notice that the integral function in Eq. 2.35 is actually a periodic function, and thus it does not make any sense to calculate this integration up to infinite time range. However, in real materials, there always exist various external scattering mechanisms like defects and additional interaction with the environment, both of which have not been considered in the dimer model for the charge transfer rate here. Therefore the time range for integration in Eq. 2.35 is always limited. In practice, when the electron–phonon coupling is large and/or the temperature is high enough, the integral function in Eq. 2.35 can decay very quickly and remain negligible for a long time which is comparable with the period of the function. Therefore it is a quite reasonable to calculate the integration just within the first period. If it is not the case, namely, the integral function oscillates with time and does not really decay to very small values, one needs to make a kind of cutoff to the integral time. One advisable approach is to choose the most important mode for the charge transfer process, i.e., the mode with the most significant Huang–Rhys factor, and apply the short-time approximation $\exp(it\omega_j) \approx 1 + it\omega_j + (it\omega_j)^2/2$, where the last term provides an overall decay factor in the integrand and guarantees the convergence for Eq. 2.35.

2.5.3 From Fermi's Golden Rule to Marcus Rate

In the strong coupling limit (i.e., $\sum_j S_j \gg 1$) and/or the high-temperature limit ($k_B T \gg \hbar\omega_j$), we can use the short-time approximation automatically for all the vibrational modes, namely, keeping only the three leading terms in the infinite expansion $\exp(it\omega_j) = 1 + it\omega_j + (it\omega_j)^2/2 + \cdots$, and then Eq. 2.34 becomes

$$k = \frac{V^2}{\hbar^2} \int_{-\infty}^{\infty} dt \exp\left[it\left(\omega_{fi} + \sum_j S_j\omega_j\right) - \frac{t^2}{2}\sum_j S_j\omega_j^2(2n_j + 1)\right] \tag{2.36}$$

or

$$k = \frac{V^2}{\hbar^2}\sqrt{\frac{2\pi}{\sum_j S_j\omega_j^2(2n_j+1)}}\exp\left[-\frac{\left(\omega_{fi} + \sum_j S_j\omega_j\right)^2}{2\sum_j S_j\omega_j^2(2n_j+1)}\right] \tag{2.37}$$

In the high-temperature regime, we have $n_j = k_B T/\hbar\omega_j \gg 1$, and Eq. 2.37 reduces to the Marcus formula, Eq. 2.1, with $\Delta G^0 = \hbar\omega_{fi}$ and $\lambda = \sum_j S_j\hbar\omega_j$.

2.6 Application: Oligoacenes

Oligoacenes such as tetracene, rubrene, and pentacene (see Fig. 2.16) are among the most promising classes of organic semiconductors for (opto)electronic applications [40]. The planarity and rigidity of tetracene and pentacene molecules facilitate good intermolecular ordering, and their extended π-conjugation over the whole molecule enables large intermolecular electron overlap. Ruberene is a star derivative of tetracene with additional four phenyl side groups, which further enhance tight crystal packing. Here, we use tetracene and ruberene as model systems to investigate the role of nuclear tunneling effect on charge transport properties.

2.6.1 Computational Details

Transfer integrals are obtained with the direct scheme with Kohn-Sham–Fock operator described in Sect. 2.1.2.1. Normal mode analysis is performed to get all the intramolecualr vibrational frequencies and the corresponding Huang–Rhys factors. The mobility is obtained through random walk simulation with the quantum charge transfer rate in Eq. 2.35. Tetracene and Ruberene form layer-by-layer crystals. The chosen hopping pathways within the layer are shown in Fig. 2.17.

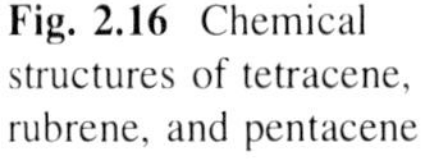

Fig. 2.16 Chemical structures of tetracene, rubrene, and pentacene

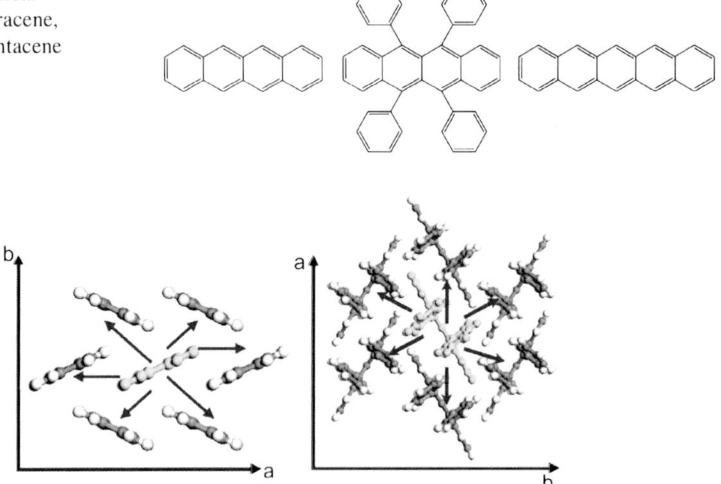

Fig. 2.17 Chosen intra-layer pathways for tetracene crystal (*left*) and ruberene crystal (*right*). Reprinted from Ref. [39] by permission of APS

2.6.2 Results and Discussion

2.6.2.1 Transfer Integral

The largest transfer integral calculated for rubrene comes from the *a* direction (see Fig. 2.17), 102.4 meV, which is much larger than that for tetracene which is about 40 meV. This can be understood from their molecular packings. Both rubrene and tetracene have a herringbone motif in the *ab* plane where the most significant electronic couplings are found. However, due to the phenyl side groups, the long molecule axes lie in the *ab* plane in rubrene, while in tetracene, they come out of that plane. This modulation leads to no short-axis displacement along the *a* direction and cofacial π-stack with some long-axis displacement.

2.6.2.2 Normal Mode Analysis

The contribution of individual vibrational modes to the total reorganization energy is shown in Fig. 2.18. For tetracene, it is found that high-frequency C–C bond stretching modes present dominant electron–phonon couplings. And rubrene differs strongly with tetracene in the low-frequency region due to the twisting motions of the four phenyl groups being strongly coupled with the charge transfer process, similar with the results for oligothiophenes shown in Fig. 2.14. As a result, the reorganization energy of rubrene (150 meV) is much larger than that of tetracene (105 meV).

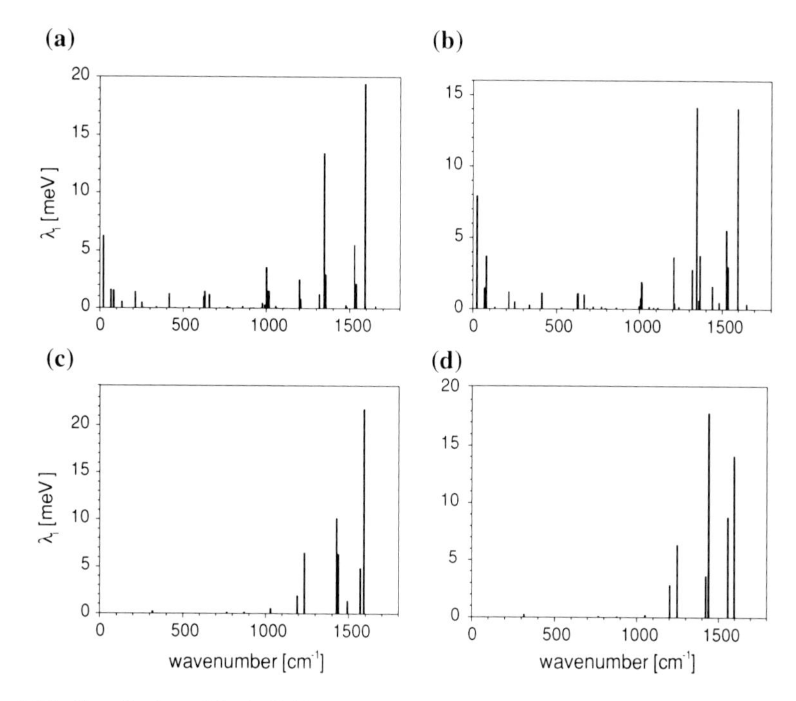

Fig. 2.18 Contribution of the individual vibrational modes to the relaxation energies for neutral and cationic molecules, (**a**) neutral rubrene, (**b**) cationic rubrene, (**c**) neutral tetracene, and (**d**) cationic tetracene. Reprinted from Ref. [39] by permission of APS

2.6.2.3 Temperature Dependence of Mobility

We calculate the temperature-dependent charge transfer rate with both the Marcus theory and the Fermi's golden rule (see Fig. 2.19a). It is found that the quantum transfer rate is nonzero and almost constant below 10 K due to the quantum tunneling nature of the nuclear vibrations at low temperatures. It then increases with temperature and reaches a maximum at about 130 K and finally decreases gradually. This behavior is significantly different with the classical Marcus rate. For the transition point of the charge transfer rate at higher temperatures, which corresponds to the thermal activation over a barrier, the classical Marcus theory gives $\lambda/4 = 37.5$ meV $\sim$ 435 K, while the nuclear tunneling reduces this height to only 130 K. The corresponding temperature dependence of mobility for rubrene is shown in Fig. 2.19b. The mobility decreases rapidly from 1 to 10 K, then increases slowly until 30 K, and decreases slowly again at higher temperatures. These can be fully explained by the general $k(T)/T$ behavior within the hopping picture. For tetracene, the charge transfer rate remains constant up to room temperature (see Fig. 2.20) since the contributed vibrations are all high-frequency modes and the nuclear tunneling effect is extremely strong. Accordingly, the mobility decreases with temperature throughout the whole temperature range, showing a band-like behavior.

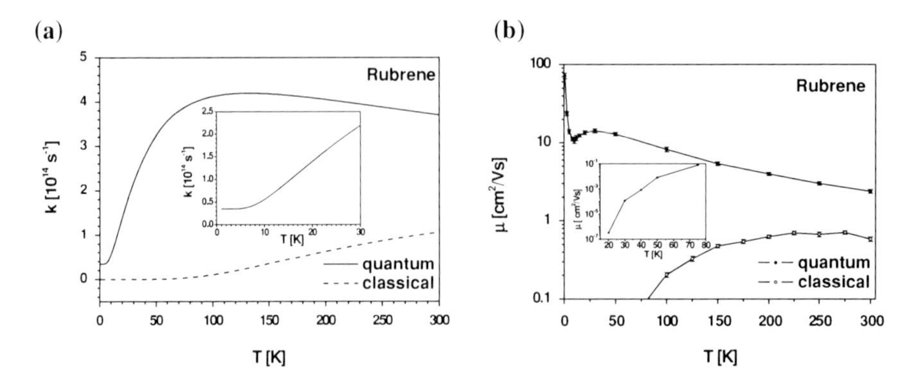

Fig. 2.19 **a** Hole transfer rate as a function of temperature for the rubrene dimer with the largest transfer rate. The insets shows quantum CT rate below 5 K; **b** 3D averaged hole mobilities as a function of temperature. The inset shows the mobility from Marcus theory in low temperature. Reprinted from Ref. [39] by permission of APS

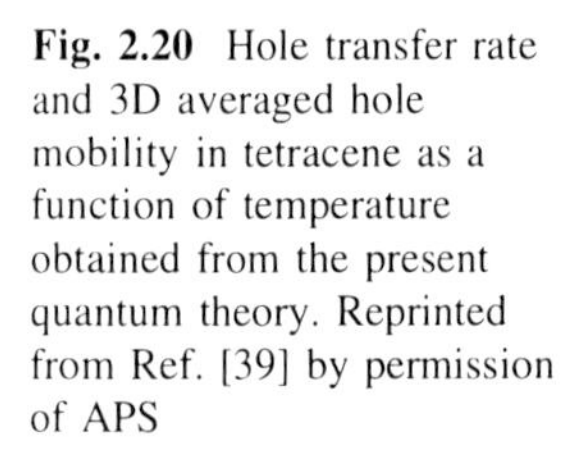

Fig. 2.20 Hole transfer rate and 3D averaged hole mobility in tetracene as a function of temperature obtained from the present quantum theory. Reprinted from Ref. [39] by permission of APS

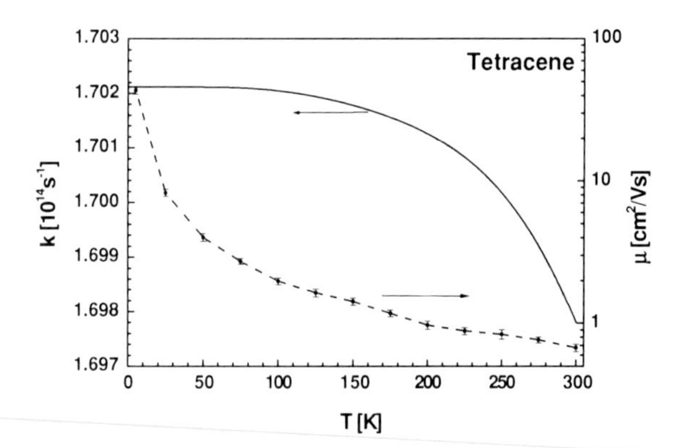

2.6.2.4 Anisotropy of Mobility

Notice that the results shown above are averaged over all directions in 3D and thus are isotropic mobilities. Considering that the real molecular crystals are highly anisotropic, the comparison between the calculated data here and the experiment along each direction is not reasonable. To investigate the strength of anisotropy, one can also perform the 2D-averaged simulations within the layer plane of the crystals. It is found that the mobility is reduced by a factor of 2.5–2.8 for tetracene and rubrene from 2D to 3D structures due to the weaker interlayer couplings.

2.7 Incorporate Dynamic Disorder Effect in the Hopping Picture

2.7.1 Two-Step Approach

In both the general hopping description described in Sect. 2.1 and the improved approach with nuclear tunneling effect introduced in Sect. 2.5, only the local electron–phonon couplings are considered, namely the site energies are modulated by nuclear vibrations while the intermolecular electronic couplings are kept fixed. However, at room temperature, it is obvious that the relative orientation of molecules fluctuates all the time since the intermolecular interaction is van der Waals-type weak, and thus the transfer integrals are also strongly modulated by nuclear motions. We notice that the nonlocal electron–phonon couplings are dominated by low-frequency intermolecular modes [41] with period (about 600 fs) much larger than the time of a single charge transfer process (a few to tens of femtoseconds) for good organic semiconductors [42]. Therefore one can perform a two-step approach to include the nonlocal electron–phonon couplings inside the hopping picture, namely, the transfer integrals are kept constant during the charge transfer processes and they are updated after each hopping step.

2.7.2 Multi-Scale Simulation

It is mentioned above that the nuclear vibrations which are important for modulating the transfer integrals are low-frequency modes, thus it is quite straightforward to introduce molecular dynamics (MD) to describe their classical manner. One can build a supercell containing enough number of investigated molecules, and perform MD simulations to get the trajectory of the nuclear dynamics. At each snapshot, the intermolecular transfer integrals can be calculated with the dimer methods quantum chemically described in Sect. 2.1.2. To get real time evolution of the transfer integrals, a discrete Fourier transformation is performed to these discrete data points:

$$V_{mn}(t) = \langle V_{mn} \rangle + \sum_{k=0}^{N/2} \mathrm{ReV}_k \cos(\omega_k t + \phi_0) + \sum_{k=0}^{N/2} \mathrm{ImV}_k \sin(\omega_k t + \phi_0) \quad (2.38)$$

Here, N is the total number of MD snapshots, ReV and ImV are the amplitudes of cosine and sine basis functions, on the basis of which the contributions of different phonons to the transfer integral fluctuation can be achieved. The same type of molecular dimers in the crystal should have the same Fourier coefficients with different phase factors ϕ_0. The phase factors can be chosen randomly because there's hardly any fluctuation correlation between transfer integrals of different pairs [42]. Therefore, one can deal with several typical dimers to get the Fourier coefficients, and the time-dependent transfer integrals between all molecular dimers can be realized according to Eq. 2.38 with different phase factors. Random walk Monte Carlo simulation technique can be carried out to investigate the charge carrier mobility after slight modifications. For example, initially for each molecular dimer, a phase factor ϕ_0 is chosen randomly as $r\omega_k t_{\mathrm{simu}}$, where r is uniformly distributed in [0, 1] and t_{simu} is the total MD simulation time. And the transfer integrals are updated using Eq. 2.38 for the new time after each hopping step.

2.8 Application: Pentacene

The chemical structure of pentacene is already shown previously in Fig. 2.16. The thin-film phase of pentacene is a substrate-induced polymorph, commonly existing in pentacene thin-film transistors with hole mobility exceeding 5.0 cm^2/Vs [43], and has thus received a lot of attention. Here, we perform a multi-scale approach proposed above, namely, MD simulations to achieve the time evolution of molecule geometries, quantum chemical calculations for the transfer integrals at each MD snapshot, and Monte Carlo method to simulate charge carrier diffusion, to study the charge transport mechanism in the thin-film phase of pentacene crystal.

2.8.1 Computational Details

We choose a $3 \times 3 \times 3$ supercell for molecular dynamics based on the experimental crystal structure (see Fig. 2.21) [44]. The MD simulation with fixed lattice constants is carried out at five constant temperatures, i.e., 100, 150, 200, 250, and 300 K with COMPASS force field within the Materials Studio package [45]. The simulation time is set to be 100 ps with a time step of 2 fs, and the dynamic trajectories are extracted every 30 fs after thermal equilibration of 40 ps with a total number of 2,000 snapshots. Within one layer, each molecule has six nearest neighbors. From the symmetry, we only calculate the transfer integrals for typical molecular dimers A, B, and C (see Fig. 2.21). The transfer integrals at each snapshot are calculated with site-energy correction method in Sect. 2.1.2.2 at the PW91PW91/6-31G* level within Gaussian 03 package [23]. The simulation time for a single Monte Carlo is 10 ps and 5,000 simulations are performed to get the carrier mobility.

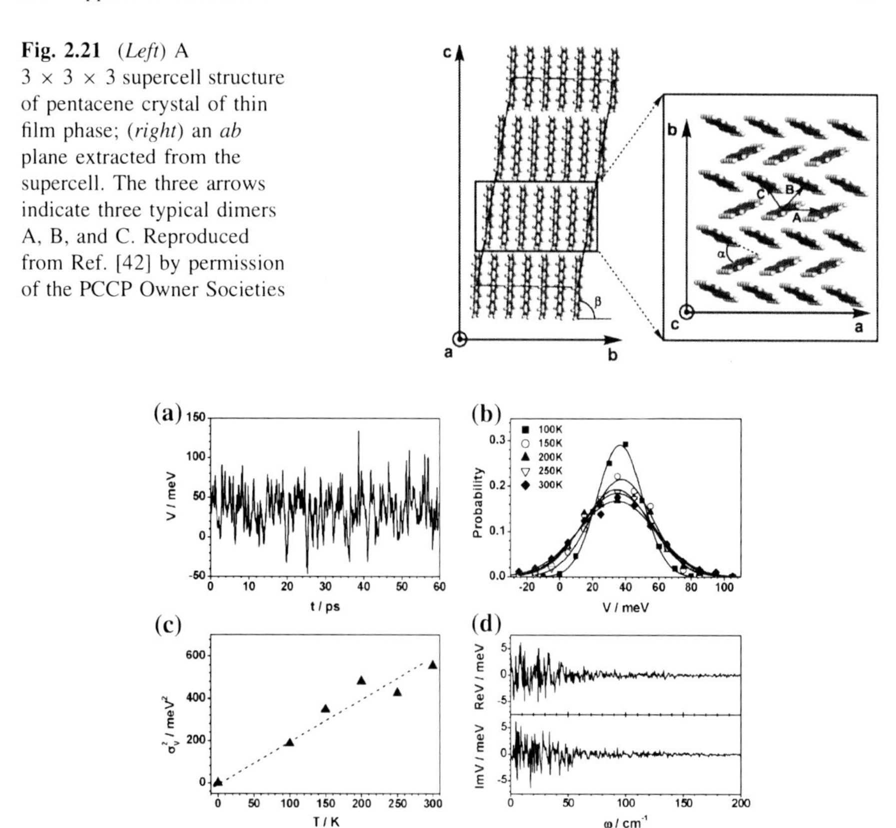

Fig. 2.21 (*Left*) A 3 × 3 × 3 supercell structure of pentacene crystal of thin film phase; (*right*) an *ab* plane extracted from the supercell. The three arrows indicate three typical dimers A, B, and C. Reproduced from Ref. [42] by permission of the PCCP Owner Societies

Fig. 2.22 **a** Thermal fluctuation of the transfer integral (dimer A) at 300 K; (**b**) distribution of the transfer integrals at different temperatures; **c** square of the standard deviation of transfer integral versus temperature, where σ_V is set to zero at zero temperature in classical limit; **d** Fourier transformation of thermal deviation amplitude at 300 K. Reproduced from Ref. [42] by permission of the PCCP Owner Societies

2.8.2 Results and Discussion

2.8.2.1 Transfer Integral Fluctuation

A typical calculated time evolution of the transfer integral is shown in Fig. 2.22a. The thermal fluctuation is of the same order of magnitude as the average value, which agrees with observation in other polymorph of pentacene crystals [41]. They follow the Gaussian distributions with almost temperature independent mean

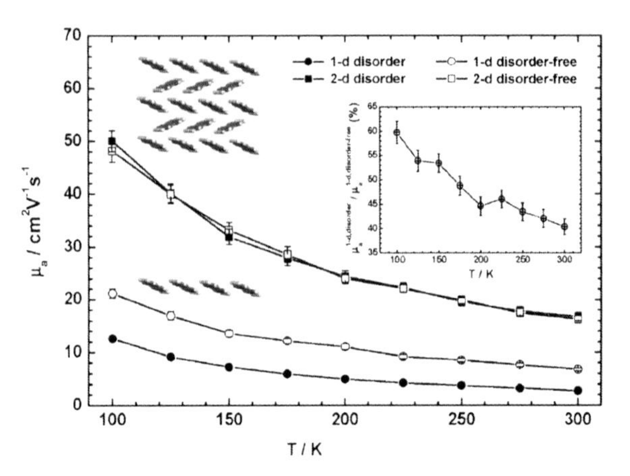

Fig. 2.23 Temperature-dependent hole mobility along the *a* axis with and without dynamic disorder in 1D (*circles*) and 2D (*squares*). Ratio between the mobility with and without dynamic disorder in 1D is shown in the inset. Reproduced from Ref. [42] by permission of the PCCP Owner Societies

values (see Fig. 2.22b), and the square of the standard deviation σ_V is a linear function of temperature (see Fig. 2.22c). This can be understood by the combined effects of Boltzmann distribution of intermolecular distances and the widely assumed linear electron–phonon coupling. Accordingly, the Fourier coefficients, ReV and ImV, should also follow the $T^{0.5}$ law. From Fig. 2.22d, we reproduce that the major contribution to the transfer integral fluctuation comes from low-frequency modes (<50 cm^{-1}), belonging to intermolecular vibrations.

2.8.2.2 Mobility in 1D and 2D Cases

We investigate the 1D (i.e., a molecular chain along *a* direction of the crystal) and 2D (i.e., the molecular layer within the *ab* plane) temperature-dependent hole mobility both with and without thermal fluctuation of the transfer integrals (see Fig. 2.23). For the 1D case, due to the fluctuating nature of the transfer integrals, the charge transfer rates between some of the molecular dimers become less than those at the equilibrium geometry and the charge becomes oscillating between dimers with larger charge transfer rate, becoming bottle necks for charge transport. It is noted that even at low temperature (100 K), the disorder effect is remarkable. This is due to the fact that the dominant intermolecular mode is around 50 cm^{-1}, which can be converted to about 72 K. The ratio between the simulated mobility with and without dynamic disorder decreases with temperature due to the larger fluctuation of transfer integrals and thus more pronounced "bottleneck effect" at higher temperatures. For the 2D case, it is interesting to see that the temperature dependence of mobility does not depend on the dynamic disorder of the transfer integrals. In 2D systems, there are much more hopping pathways than in 1D case. If the transfer integral of one path is small, the hole can always choose another pathway with larger transfer integrals, thus the mobility is less affected. Note that here the charge transfer rates are temperature independent below room temperature, similar with the case of tetracene shown in Fig. 2.20. As a result, the "band-

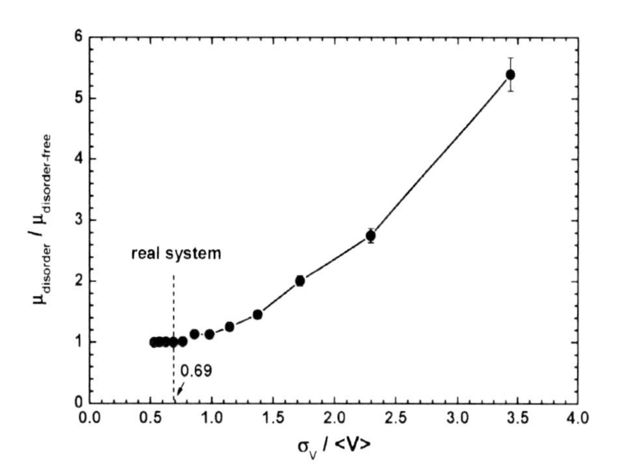

Fig. 2.24 Ratio of the hole mobility along a axis in 2D with dynamic disorder with respect to that of disorder-free, as a function of σ_V over the mean transfer integral $\langle V \rangle$. Here, σ_V is kept at the value in 300 K, while $\langle V \rangle$ is varied. The phonon-assisted current starts at about $\sigma_V / \langle V \rangle = 1$. Reproduced from Ref. [42] by permission of the PCCP Owner Societies

like" behavior of temperature dependence of mobility is solely a nuclear tunneling effect. In an extreme case, when the fluctuation in transfer integral is larger than its average, the dynamic disorder can even increase the hole mobility for 2D (see Fig. 2.24). This indicates an intrinsic transition to the phonon-assisted transport by dynamic disorder. More details about the role of such non-local electron–phonon couplings will be talked about in Chap. 3 within the polaron picture.

2.9 Conclusion

In this chapter, we have talked about the general methodology of the hopping mechanism as well as two improvements incorporating the nuclear tunneling effect from the high-frequency modes for the local electron–phonon couplings and the dynamic disorder effect from the low-frequency modes for the nonlocal electron–phonon couplings. Applications of these approaches have been performed to various star organic semiconducting materials, e.g., the siloles, triphenylamines, oligothiophenes, and oligoacenes. Several clear conclusions can be drawn: (1) The transfer integrals between neighboring molecules and the reorganization energy during the charge transfer processes are key to determine the charge transport efficiencies in molecular crystals. The former requires tight crystal packing and nice intermolecular matching between frontier molecular orbitals, while the latter favors long conjugated, planer, and rigid molecules with less flexible degrees of freedom. (2) At low temperatures, the nuclear tunneling effect can be very important for charge transport. Because of this, the "band-like" temperature dependence of mobility is possible within the hopping picture when the major contribution comes from the high-frequency modes. We note that when the external reorganization, which is mostly of low-frequency environment mechanisms, is very strong, the overall temperature dependence may be changed completely, and then the high-frequency modes will only generally reduce the mobility, without any impact on temperature dependence. (3) The dynamic disorder coming from the thermal fluctuation of the transfer integrals is very important for one-dimensional systems. For higher dimensions, the effect becomes significant only when the fluctuation is larger than the mean transfer integral itself. Therefore this effect should be considered for loosed packing crystals and high temperatures. There, a phonon-assisted term will be added to the temperature dependence of mobility, and can be of great importance to the overall transport behavior. In Chap. 3, we will talk about the Holstein–Peierls polaron model, which is principally more general than the present hopping model. More basic understandings concerning the role of different electron–phonon couplings on the temperature dependence of mobility will be revealed.

References

1. R.A. Marcus, Rev. Mod. Phys. **65**, 599 (1993)
2. T. Fujita, H. Nakai, H. Nakatsuji, J. Chem. Phys. **104**, 2410 (1996)
3. E.F. Valeev, V. Coropceanu, D.A. da Silva Filho, S. Salman, J.-L. Brédas, J. Am. Chem. Soc. **128**, 9882 (2006)
4. K. Hannewald, V.M. Stojanović, J.M.T. Schellekens, P.A. Bobbert, G. Kresse, J. Hafner, Phys. Rev. B **69**, 075211 (2004)
5. A. Troisi, G. Orlandi, J. Phys. Chem. B **106**, 2093 (2002)
6. J.M. André, J. Delhalle, J.-L. Brédas, *Quantum Chemistry Aided Design of Organic Polymers, An Introduction to the Quantum Chemistry of Polymers and its Applications* (World Scientific, Singapore, 1991)
7. S.W. Yin, Y.P. Yi, Q.X. Li, G. Yu, Y.Q. Liu, Z.G. Shuai, J. Phys. Chem. A **110**, 7138 (2006)
8. J.-L. Brédas, D. Beljonne, V. Coropceanu, J. Cornil, Chem. Rev. **104**, 4971 (2004)
9. X.D. Yang, L.J. Wang, C.L. Wang, W. Long, Z. Shuai, Chem. Mater. **20**, 3205 (2008)
10. A. Einstein, Ann. Phys. **17**, 549 (1905)
11. M. von Smoluchowski, Ann. Phys. **21**, 756 (1906)
12. L.B. Schein, A.R. McGhie, Phys. Rev. **20**, 1631 (1979)
13. W.Q. Deng, W.A. Goddard III, J. Phys. Chem. B **108**, 8624 (2004)
14. L.J. Wang, G.J. Nan, X.D. Yang, Q. Peng, Q.K. Li, Z. Shuai, Chem. Soc. Rev. **39**, 423 (2010)
15. J.D. Luo, Z. Xie, J.W.Y. Lam, L. Cheng, H. Chen, C. Qiu, H.S. Kwok, X. Zhan, Y. Liu, D. Zhu, B.Z. Tang, Chem. Commun. **18**, 1740 (2001)
16. B.Z. Tang, X. Zhan, G. Yu, P.P.S. Lee, Y. Liu, D. Zhu, J. Mater. Chem. **11**, 2874 (2001)
17. L.C. Palilis, A.J. Makinen, M. Uchida, Z.H. Kafafi, Appl. Phys. Lett. **82**, 2209 (2003)
18. H. Murata, Z.H. Kafafi, Appl. Phys. Lett. **80**, 189 (2002)
19. M.A. Baldo, D.F. O'Brien, Y. You, A. Shoustikov, S. Silbey, M.E. Thompson, S.R. Forrest, Nature (London) **395**, 151 (1998)
20. M.A. Baldo, M.E. Thompson, S.R. Forrest, Nature (London) **750**, 403 (2000)
21. J.W. Chen, C.C.W. Law, J.W.Y. Lam, Y.P. Dong, S.M.F. Lo, I.D. Williams, D.B. Zhu, B.Z. Tang, Chem. Mater. **15**, 1535 (2003)
22. G. Yu, S.W. Yin, Y.Q. Liu, J.S. Chen, X.J. Xu, X.B. Sun, D.G. Ma, X.W. Zhan, Q. Peng, Z. Shuai, B.Z. Tang, D.B. Zhu, W.H. Fang, Y. Luo, J. Am. Chem. Soc. **127**, 6335 (2005)
23. M.J. Frisch et al., *Gaussian 03, revision A. 1* (Gaussian, Inc, Pittsburgh, 2003)
24. D. Beljonne, A.J. Ye, Z.G. Shuai, J.L. Brédas, Adv. Funct. Mater. **14**, 684 (2004)
25. Y. Shirota, J. Mater. Chem. **10**, 1 (2000)

26. Y. Shirota, J. Mater. Chem. **15**, 75 (2005)
27. A. Cravino, S. Roquet, O. Aleveque, P. Leriche, P. Frere, J. Roncali, Chem. Mater. **18**, 2584 (2006)
28. T.P.I. Saragi, T.F. Lieker, J. Salbeck, Adv. Funct. Mater. **16**, 966 (2006)
29. M. Sonntag, K. Kreger, D. Hanft, P. Strohriegl, S. Setayesh, D. de Leeuw, Chem. Mater. **17**, 3031 (2005)
30. Y.B. Song, C.A. Di, X.D. Yang, S.P. Li, W. Xu, Y.Q. Liu, L.M. Yang, Z. Shuai, D.Q. Zhang, D.B. Zhu, J. Am. Chem. Soc. **128**, 15940 (2006)
31. X.D. Yang, Q.K. Li, Z.G. Shuai, Nanotechnology **18**, 424029 (2007)
32. D. Fichou, *Handbook of Oligo- and Polythiophenes* (Wiley-VCH, New York, 1999)
33. W.A. Schoonveld, J. Wildeman, D. Fichou, P.A. Bobbert, B.J. van Wees, T.M. Klapwikj, Nature (London) **404**, 977 (2000)
34. G. Barbarella, M. Zambianchi, A. Bongini, L. Antolini, Adv. Mater. **5**, 834 (1993)
35. P. Weber, J.R. Reimers, J. Phys. Chem. A **103**, 9830 (1999)
36. G. Zerbi, H.W. Siesler, I. Noda, M. Tasumi, S. Krimm, *Modern Polymer Spectroscopy* (Wiley, New York, 1999)
37. J. Jortner, J. Chem. Phys. **64**, 4860 (1976)
38. S.H. Lin, C.H. Chang, K.K. Liang, R. Chang, Y.J. Shiu, J.M. Zhang, T.S. Yang, M. Hayashi, F.C. Hsu, Adv. Chem. Phys. **121**, 1 (2002)
39. G.J. Nan, X.D. Yang, L.J. Wang, Z. Shuai, Y. Zhao, Phys. Rev. B **79**, 115203 (2009)
40. J.E. Anthony, Chem. Rev. **106**, 5028 (2006)
41. A. Troisi, G. Orlandi, J. Phys. Chem. A **110**, 4065 (2006)
42. L.J. Wang, Q.K. Li, Z. Shuai, L.P. Chen, Q. Shi, Phys. Chem. Chem. Phys. **12**, 3309 (2010)
43. T.W. Kelley, D.V. Muyres, P.F. Baude, T.P. Smith, T.D. Jones, Mater. Res. Soc. Symp. Proc. **771**, L6.5 (2003)
44. S. Schiefer, M. Huthm, A. Dobrinevski, B. Nickel, J. Am. Chem. Soc. **129**, 10316 (2007)
45. H. Sun, J. Phys. Chem. B **102**, 7338 (1998)

Chapter 3
Polaron Mechanism

Abstract The intrinsic charge transport in organic semiconductors is an electron–phonon interacting process. Due to the "soft" nature of organic materials, the existence of an electron can cause significant deformation of local nuclear vibrations, which moves together with the electron itself, and thus the effective diffusing quasiparticle is composed of the electron and its accompanying phonons. This is the basic idea of the polaron mechanism. In principle, it is a more general description for charge transport since it does not presume that the charge is localized within one molecule as in the hopping mechanism described in Chap. 2. In this chapter, we adopt the general Holstein-Peierls Hamiltonian coupled with first-principles calculations to investigate the fundamental aspects concerning charge transport. All kinds of electron–phonon couplings, including both local and nonlocal parts for inter- and intra-molecular vibrations, have been taken into considerations. Detailed studies are performed to study their contributions to the total electron–phonon coupling strength and the temperature dependence of mobility, especially the band-hopping crossover feature. We also investigate the pressure- and temperature-dependent crystal structure effects on the charge transport properties.

Keywords Polaron mechanism · Holstein-Peierls Hamiltonian · Electron–phonon coupling · Band-hopping crossover · Pressure and temperature dependence of mobility · Thermal expansion of lattice

In Sect. 3.1, we discuss the derivation for the mobility expression from Holstein-Peierls Hamiltonian with polaron transformation and linear response theory. Application to naphthalene crystal is presented in Sect. 3.2 to investigate the role of inter- and intra-molecular vibrations on the temperature dependence of mobility. In Sect. 3.3, the temperature dependence is improved after including the thermal expansion of the lattice. Finally, the pressure dependence of mobility is studied in Sect. 3.4.

Z. Shuai et al., *Theory of Charge Transport in Carbon Electronic Materials*,
SpringerBriefs in Molecular Science, DOI: 10.1007/978-3-642-25076-7_3,

3.1 Holstein-Peierls Model

In 1959, Holstein proposed the local electron–phonon interaction model for the small polaron transport in one-dimensional molecular crystals, and derived the analytical result by means of the perturbation theory [1]. The Holstein model provides the origin of band to hopping crossover transport behavior, and depicted a general scheme for studying charge transport in organic solids [2]. An attempt to generalize the Holstein model to the Holstein-Peierls model by including the nonlocal electron–phonon couplings was made by Munn and Silbey [3]. It is found that nonlocal couplings in general tend to increase scattering, thereby reducing band and increasing hopping contributions to the mobility. Kenkre et al. has applied the Holstein model to high dimensions, and given a unified quantitative explanation of the mobility behavior in naphthalene crystal for the first time, by assuming directionally dependent local electron–phonon coupling constants [4]. Recently, Hannewald and coauthors reexamined the charge transport in naphthalene crystal based on the Holstein-Peierls Hamiltonian with parameters calculated at the ab initio level [5]. The experimental temperature dependences for electron and hole mobility as well as the spatial anisotropy have been qualitatively reproduced by taking only three intermolecular vibrations into considerations. The process to derive the mobility formula for Holstein-Peierls polaron is technical, and has been described in detail by Hannewald et al. [6, 7]. Here we only list the most important steps during the derivations.

3.1.1 Holstein-Peierls Hamiltonian

The Holstein-Peierls Hamiltonian is composed of three parts: the electronic part (H_{e}), the phonon part (H_{p}), and the electron-coupling part ($H_{\mathrm{e-p}}$):

$$\begin{aligned} H &= H_{\mathrm{e}} + H_{\mathrm{p}} + H_{\mathrm{e-p}} \\ &= \sum_{mn} \varepsilon_{mn} a_m^+ a_n + \sum_{\lambda} \hbar\omega_\lambda \left(b_\lambda^+ b_\lambda + \frac{1}{2} \right) + \sum_{mn\lambda} \hbar\omega_\lambda g_{\lambda mn} \left(b_\lambda^+ + b_{-\lambda} \right) a_m^+ a_n \end{aligned} \tag{3.1}$$

Here, the operator $a_m^{(+)}$ represents annihilating (creating) an electron at the lattice site m with energy ε_{mm} and $b_\lambda^{(+)}$ represents annihilating (creating) a phonon with frequency ω_λ. ε_{mn} is the transfer integral between molecules m and n. $g_{\lambda mn}$ is the local ($m = n$, Holstein model) or nonlocal ($m \neq n$, Peierls model) dimensionless electron–phonon coupling constant which characterizes the interaction strength between phonon λ and the onsite energy ε_{mm} or the transfer integral ε_{mn}.

3.1.2 Polaron Transformation

In order to get an effective Hamiltonian to characterize the properties of the polaron, one generally starts from the following canonical transformation [6]

$$H \to \tilde{H} = e^S H e^{-S} \tag{3.2}$$

with

$$S = \sum_{\lambda mn} g_{\lambda mn} (b_\lambda^+ - b_{-\lambda}) a_m^+ a_n \equiv \sum_{mn} C_{mn} a_m^+ a_n \tag{3.3}$$

After insertion of several factors $1 = e^{-S} e^S$, it is easy to check that

$$\tilde{H}\left(a_m^{(+)}, b_\lambda^{(+)}\right) = H\left(\tilde{a}_m^{(+)}, \tilde{b}_\lambda^{(+)}\right) \tag{3.4}$$

where

$$\tilde{a}_m = e^S a_m e^{-S} = \sum_n \left(e^{-C}\right)_{mn} a_n \tag{3.5}$$

$$\tilde{a}_m^+ = e^S a_m^+ e^{-S} = \sum_n a_n^+ \left(e^C\right)_{nm} \tag{3.6}$$

$$\tilde{b}_\lambda = e^S b_\lambda e^{-S} = b_\lambda - \sum_{mn} g_{\lambda mn} a_m^+ a_n \tag{3.7}$$

$$\tilde{b}_\lambda^+ = e^S b_\lambda^+ e^{-S} = b_\lambda^+ - \sum_{mn} g_{-\lambda mn} a_n^+ a_m = b_\lambda^+ - \sum_{mn} g_{-\lambda mn} a_m^+ a_n \tag{3.8}$$

Then the transformed Hamiltonian becomes [6]

$$\tilde{H} = e^S H e^{-S} = \sum_{mn} \tilde{E}_{mn} a_m^+ a_n + \sum_\lambda \hbar\omega_\lambda \left(b_\lambda^+ b_\lambda + \frac{1}{2}\right) \tag{3.9}$$

where

$$\tilde{E}_{mn} = \left(e^C E e^{-C}\right)_{mn} \tag{3.10}$$

$$E_{mn} = \varepsilon_{mn} - \sum_\lambda \hbar\omega_\lambda (g_\lambda g_{-\lambda})_{mn} \tag{3.11}$$

By means of the Baker-Campbell-Hausdorff theorem,

$$\tilde{A} = e^C A e^{-C} = \sum_{k=0}^{\infty} \frac{1}{k!} \underbrace{[C, [C, \ldots, [C, A] \ldots]]}_{k \text{ commutators}} \tag{3.12}$$

where A is any operator, Eq. 3.10 becomes [6]

$$\tilde{E}_{mn} = \sum_{k=0}^{\infty} \frac{1}{k!} \sum_{\lambda_1 \ldots \lambda_k} [g_{\lambda_1}, [g_{-\lambda_1}, \ldots [g_{\lambda_k}, [g_{-\lambda_k}, E]] \ldots]]_{mn} \left(b_{\lambda_1}^+ - b_{-\lambda_1}\right) \ldots \left(b_{\lambda_k}^+ - b_{-\lambda_k}\right) \tag{3.13}$$

3.1.3 Some Major Approximations

When comparing Eqs. 3.1 and 3.9, it seems that the electron–phonon coupling term has been removed in the transformed Hamiltonian. Actually this is not the case. The phonon operators still exist in the electronic part of Eq. 3.9 as seen in Eq. 3.13. Normally, it is assumed that the thermal average of Eq. 3.13 can be used to average out the phonon operators and get the approximated polaron Hamiltonian, where the polaron term and the remaining phonon term are completely decoupled. Analytically, we can get [6]

$$\langle \tilde{E}_{mn} \rangle = \sum_{k=0}^{\infty} \frac{1}{k!}\left(-\frac{1}{2}\right)^k \sum_{\lambda_1 \ldots \lambda_k} [g_{\lambda_1}, [g_{-\lambda_1}, \ldots [g_{\lambda_k}, [g_{-\lambda_k}, E]] \ldots]]_{mn}(1+2n_{\lambda_1})\ldots(1+2n_{\lambda_k}) \tag{3.14}$$

where $n_\lambda = 1/(\exp(\hbar\omega_\lambda/k_BT)-1)$ is the phonon occupation number of phonon λ. If we just take the most important contributions to Eq. 3.14, it has a very simple and clear form [7]:

$$\langle \tilde{E}_{mm} \rangle = E_{mm} \tag{3.15}$$

$$\langle \tilde{E}_{mn} \rangle = E_{mn} \exp\left(-\frac{1}{2}\sum_{\lambda} G_{\lambda mn}(1 + 2n_\lambda)\right) \tag{3.16}$$

where

$$G_{\lambda mn} = (g_{\lambda mm} - g_{\lambda nn})^2 + \sum_{k \neq m} g_{\lambda mk}^2 + \sum_{k \neq n} g_{\lambda nk}^2 \tag{3.17}$$

Finally, as discussed by Mahn [8], one can just keep the diagonal terms of thermal averaged electronic Hamiltonian, which means that the finite polaron bandwidth effect is partially neglected [7]. one can get

$$\tilde{H}' = \sum_m E_{mm} a_m^+ a_m + \sum_\lambda \hbar\omega_\lambda \left(b_\lambda^+ b_\lambda + \frac{1}{2}\right) \tag{3.18}$$

which is diagonal in both the electron and phonon operators, and thus is quite helpful for us to perform their thermal averages.

3.1.4 Linear Response Theory

Generally, the electron mobility can be obtained from the Kubo's conductivity formalism based on linear-response theory [8],

$$\mu_\alpha = \frac{1}{2k_B T e_0 n_0} \lim_{\omega \to 0} \int dt e^{i\omega t} < j_\alpha(t) j_\alpha(0) > \tag{3.19}$$

where α is a lattice direction, e_0 is the charge of an electron, n_0 is the electron density for charge transport, and the bracket indicates thermal average. The current operator, j, is defined as the time derivative of the polarization operator, P,

$$j_\alpha = \frac{dP_\alpha}{dt} = \frac{1}{i\hbar}[P_\alpha, H] \tag{3.20}$$

Within the tight-binding formalism, we have $P_\alpha = e_0 \sum_m R_{\alpha m} a_m^+ a_m$. Considering the Holstein-Peierls Hamiltonian, Eq. 3.1, we have

$$j_\alpha = \frac{e_0}{i\hbar} \sum_{mn} (R_{\alpha m} - R_{\alpha n}) E_{mn} a_m^+ a_n = \frac{e_0}{i\hbar} \sum_{mn} [R_\alpha, E]_{mn} a_m^+ a_n \tag{3.21}$$

where the matrix notation R_α only has diagonal terms $R_{\alpha m}$ expressing the coordinate of lattice site m along α axis. Similar with the transformation of the Hamiltonian, j can be also transformed into

$$\tilde{j}_\alpha = \frac{e_0}{i\hbar} \sum_{mn} \left(e^C [R_\alpha, E] e^{-C}\right)_{mn} a_m^+ a_n \tag{3.22}$$

At this point, the current–current correlation function in Eq. 3.19 can be calculated with

$$\langle j_\alpha(t) j_\alpha(0) \rangle \equiv \left\langle e^{(i/\hbar)Ht} j_\alpha e^{-(i/\hbar)Ht} j_\alpha \right\rangle_H = \left\langle e^{(i/\hbar)\tilde{H}t} \tilde{j}_\alpha e^{-(i/\hbar)\tilde{H}t} \tilde{j}_\alpha \right\rangle_{\tilde{H}} \tag{3.23}$$

Note that the thermal average is based on different Hamiltonians agreeing with that for the time-dependent current operator. In Eq. 3.23, the time-dependent current operator is difficult to evaluate analytically when the Hamiltonian is not diagonal. If the approximated transformed Hamiltonian, Eq. 3.18, is used, namely,

$$\langle j_\alpha(t) j_\alpha(0) \rangle \approx \left\langle e^{(i/\hbar)\tilde{H}'t} \tilde{j}_\alpha e^{-(i/\hbar)\tilde{H}'t} \tilde{j}_\alpha \right\rangle_{\tilde{H}'} \tag{3.24}$$

then the problem becomes much easier, since we have the identity

$$e^{(i/\hbar)\tilde{H}'t} f\left(a_m^+, a_n, b_\lambda^+, b_{\lambda'}\right) e^{-(i/\hbar)\tilde{H}'t} = f\left(a_m^+ e^{(i/\hbar)E_{mm}t}, a_n e^{-(i/\hbar)E_{nn}t}, b_\lambda^+ e^{i\omega_\lambda t}, b_{\lambda'} e^{-i\omega_{\lambda'} t}\right) \tag{3.25}$$

for any operator function f.

3.1.5 Polaron Mobility Formula

After evaluating the thermal average in Eq. 3.24 together with Eq. 3.19, we can finally get the mobility formula [9]:

$$\mu_\alpha(T) = \frac{e_0}{2k_BT\hbar^2}\sum_{n\neq m} R^2_{\alpha mn}\int dt J^2 e^{-2\Sigma_\lambda G_\lambda[1+2n_\lambda-\Phi_\lambda(t)]}e^{-\Gamma^2t^2} \tag{3.26}$$

where $R_{\alpha mn}$ is the distance between lattice sites m and n in the αth direction, $J^2 = (\varepsilon_{mn}-\Delta_{mn})^2 + \sum_q(\hbar\omega_q g_{qmn})^2\Phi_q(t)/2$, the second term of which is an inherent phonon-assisted contribution originated from nonlocal electron–phonon couplings, $\Delta_{mn} = \sum_\lambda \hbar\omega_\lambda[g_{\lambda mn}(g_{\lambda mm} + g_{\lambda nn}) + \sum_{k\neq m,n} g_{\lambda mk}g_{\lambda kn}]/2$, $G_\lambda = g^2_{\lambda mm} + \sum_{k\neq m,n} g^2_{\lambda mk}/2$ is the effective coupling constant of phonon mode λ, which includes both the local and the nonlocal parts, $n_\lambda = 1/(\exp(\hbar\omega_\lambda/k_BT)-1)$ denotes the phonon occupation number, $\Phi_q(t) = (1 + n_q)\exp(-i\omega_q t) + n_q\exp(i\omega_q t)$ describes incoherent scattering events caused by phonon number changes, and Γ is a phenomenological parameter for inhomogeneous line broadening.

3.1.6 Decomposition of Mobility Contributions

As discussed above, J^2 contains a normal intermolecular electronic coupling term and a phonon-assisted term induced by nonlocal electron–phonon couplings, and thus Eq. 3.26 can be naturally decomposed as [9]

$$\mu_\alpha(T) = A_\alpha f(T) + \sum_q B_{\alpha q}h_q(T) \tag{3.27}$$

where

$$A_\alpha \equiv \frac{e_0}{2k_B\hbar^2}\sum_{n\neq m} R^2_{\alpha mn}(\varepsilon_{mn} - \Delta_{mn})^2 \tag{3.28}$$

$$B_{\alpha q} \equiv \frac{e_0}{2k_B\hbar^2}\sum_{n\neq m} R^2_{\alpha mn}\frac{1}{2}\hbar^2\omega_q^2 g^2_{qmn} \tag{3.29}$$

$$f(T) \equiv \frac{1}{T}\int dt e^{\Sigma_\lambda -2G_\lambda(1+2n_\lambda(T))(1-\cos\omega_\lambda t)-\Gamma^2t^2}\cos\left(\sum_\lambda 2G_\lambda\sin\omega_\lambda t\right) \tag{3.30}$$

$$h_q(t) \equiv \frac{1}{T}(1+n_q(T))\int dt e^{\Sigma_\lambda -2G_\lambda(1+2n_\lambda(T))(1-\cos\omega_\lambda t)-\Gamma^2t^2}\cos\left(\sum_\lambda 2G_\lambda\sin\omega_\lambda t + \omega_q t\right) + \frac{1}{T}n_q(T)\int dt e^{\Sigma_\lambda -2G_\lambda(1+2n_\lambda(T))(1-\cos\omega_\lambda t)-\Gamma^2t^2}\cos\left(\sum_\lambda 2G_\lambda\sin\omega_\lambda t - \omega_q t\right) \tag{3.31}$$

We notice that both A_α and $B_{\alpha q}$ are constants relying only on the intrinsic parameters of the investigated material as given in the Hamiltonian, Eq. 3.1, while

$f(T)$ and $h_q(T)$ are temperature-dependent functions. If the temperature dependence of $f(T)$ and $h_q(T)$ are different, then the temperature dependence of the mobility will be determined by the relative magnitudes of A_α and $B_{\alpha q}$, which may be different in different lattice directions.

3.1.7 *Decomposition of Electron–Phonon Coupling Contributions*

In Eq. 3.26, the effective coupling constants, $\{G_\lambda\}$, are central to determine the magnitude of the charge carrier mobility. In order to characterize the contributions from each mode, we define the total effective electron–phonon coupling constant as the sum of the effective coupling constants of all the phonons:

$$G_{\text{tot}} = \sum_{\lambda} G_\lambda \tag{3.32}$$

Since each G_λ has both local and nonlocal contributions, we also decompose G_{tot} into four parts [9]:

$$\begin{aligned} G_{\text{tot}} &= \sum_{\lambda\in\text{inter}} g_{\lambda mm}^2 + \frac{1}{2}\sum_{k\neq m,\lambda\in\text{inter}} g_{\lambda mk}^2 + \sum_{\lambda\in\text{intra}} g_{\lambda mm}^2 + \frac{1}{2}\sum_{k\neq m,\lambda\in\text{intra}} g_{\lambda mk}^2 \\ &\equiv G_{\text{inter-local}} + G_{\text{inter-nonlocal}} + G_{\text{intra-local}} + G_{\text{intra-nonlocal}} \end{aligned} \tag{3.33}$$

3.2 Application: Naphthalene

Generally, there are two extreme charge transport mechanisms, i.e., band model and hopping model. At low temperature, the charge is believed to move coherently in single crystals, and the mobility decreases with temperature due to electron–phonon scatterings. At high temperature, the charge transport occurs by phonon-assisted hopping between localized states, and the mobility increases with temperature. The transition between band and hopping mechanisms in organic crystals was first observed in naphthalene crystal through the crossover in temperature dependence of mobility [10]. Thus naphthalene is normally chosen as the model systems to investigate the charge transport mechanism in molecular crystals. Here, we investigate the role of various electron–phonon couplings on charge mobility in naphthalene crystal.

3.2.1 Computational Details

In Eq. 3.26, there are six basic physical quantities which are necessary to calculate the mobility, namely, the temperature T, the site distances $\{R_{\alpha mn}\}$, the phonon frequencies $\{\omega_\lambda\}$, the transfer integrals $\{\varepsilon_{mn}\}$, the electron–phonon coupling constants $\{g_{qmn}\}$, and the inhomogeneous broadening factor Γ. At given temperature T and pressure P, $\{R_{\alpha mn}\}$ can be acquired from experimental data, and $\{\omega_\lambda\}$ can be evaluated through the diagonalization of Hessian matrix. In principle, phonon dispersion should be considered. However, due to the computational complexity, calculations are performed only at gamma-point. The transfer integrals are calculated with the band-fitting method introduced in Chap. 2. The electron–phonon coupling constants are obtained by definition, i.e., numerical differentiation of the transfer integral with respect to the phonon normal coordinate (Q_λ), $g_{\lambda mn} = 1/\left(\omega_\lambda\sqrt{\hbar\omega_\lambda}\right) \cdot \partial\varepsilon_{mn}/\partial Q_\lambda$. Finally, $\hbar\Gamma = 0.1$ meV is chosen, which is a very small value, since we focus on the conductivity in ultrapure crystals.

In practice, all calculations are performed with the Vienna ab initio simulation package, which has proved to be a powerful tool for theoretical study of periodic systems [11–13]. The lattice constants and atomic coordinates are taken from experiment [14]. The Perdew–Burke–Ernzerhof (PBE) exchange correlation (XC) functional [15] is chosen since it works well for weak intermolecular interactions in molecular crystal [16]. A $4 \times 4 \times 4$ grid in the corresponding Brillouin zone is used to fit the transfer integrals by a least-squares minimization [9].

3.2.2 Results and Discussion

3.2.2.1 Transfer Integral

The calculated band structure for the optimized naphthalene crystal is shown in Fig. 3.1. An arbitrary molecule in naphthalene crystal is chosen as reference since the site energies for type α and type β molecules in naphthalene crystal are the same due to its monoclinic $P2_1/a$ symmetry [17]. All the nearest neighboring molecules are chosen to fit the transfer integrals (see Fig. 3.2). They are denoted as $\{mn\} = \{0, a, b, c, ac, ab, abc\}$, corresponding to $R_{mn} = \{0, \pm a, \pm b, \pm c, \pm (a + c), \pm (a/2 \pm b/2), \pm (a/2 \pm b/2 + c)\}$. To validate the tight-binding band fitting method, we compare the DFT-calculated band energies with the fitted tight-binding energies for the 64 selected points in the Brillouin zone, see Fig. 3.3. It suggests that the band fitting approach works very well for naphthalene crystal and the chosen neighbors are enough to reproduce the band structure [9]. For the optimized geometry, the calculation onsite energies and transfer integrals for electron and hole transport are listed in Table 3.1. It can be easily found that the intra-layer transfer integrals (ε_a, ε_b and ε_{ab}) are of the order of several tens of meV, which is about ten times larger than the inter-layer transfer integrals (ε_c, ε_{ac} and

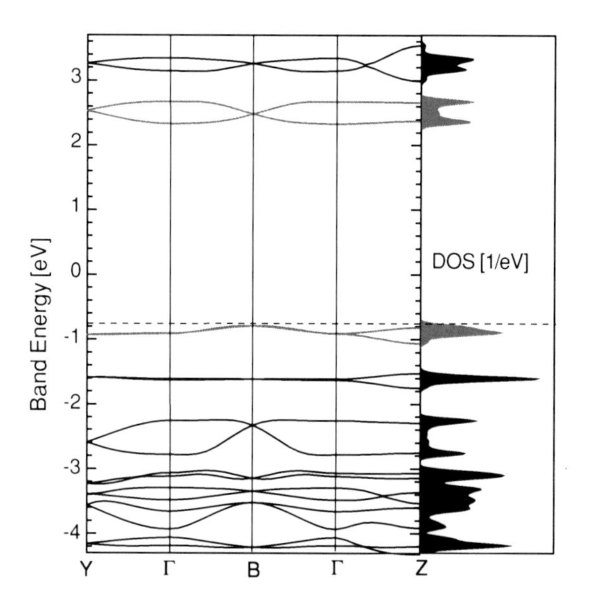

Fig. 3.1 Calculated band structure for optimized naphthalene crystal. The high symmetry points in units of $(2\pi/a, 2\pi/b, 2\pi/c)$ are $\Gamma = (0, 0, 0)$, $Y = (0.5, 0, 0)$, $B = (0, 0.5, 0)$, and $Z = (0, 0, 0.5)$. Reprinted with permission from Ref. [9]. Copyright 2007, American Institute of Physics

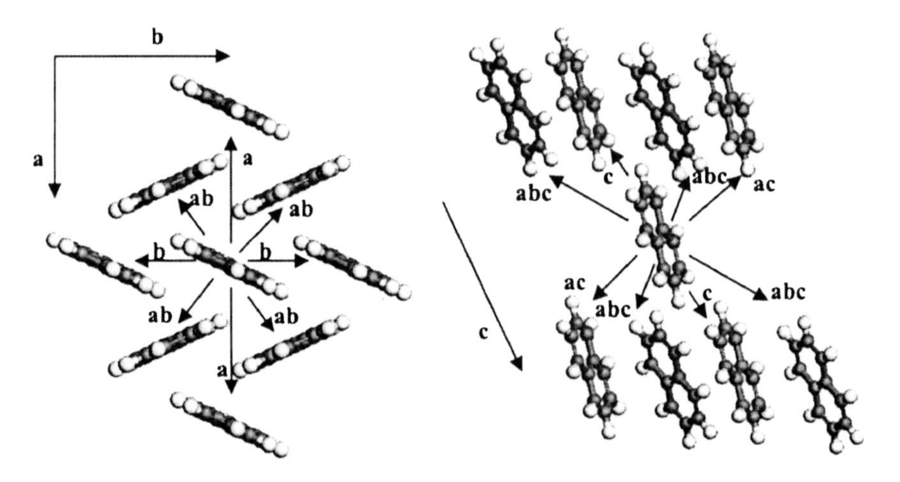

Fig. 3.2 The nearest intra-layer (*left*) and inter-layer (*right*) neighbors for of naphthalene crystal. Reprinted with permission from Ref. [20]. Copyright 2008, American Institute of Physics

ε_{abc}). This is the origin of the anisotropy of charge transport in such layer-by-layer molecular crystals.

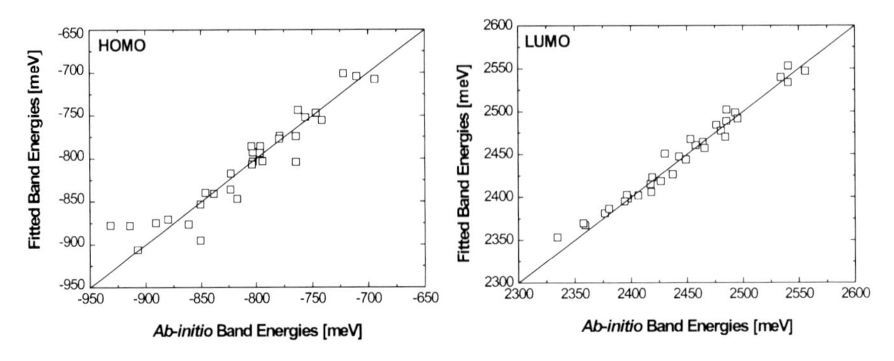

Fig. 3.3 Comparison of the DFT-calculated and the fitted tight-binding band energies for the 64 selected k points. Reprinted with permission from Ref. [9]. Copyright 2007, American Institute of Physics

Table 3.1 First principles calculated onsite energies and transfer integrals for hole and electron, in meV

	ε_0	ε_a	ε_b	ε_c	ε_{ac}	ε_{ab}	ε_{abc}
Hole	−835	−23	−42	−3	−1	22	−5
Electron	2510	7	24	−3	−1	−51	−2

3.2.2.2 Electron–Phonon Coupling Constants

The electron–phonon coupling constants are calculated for different normal modes, and the relationship between the coupling constants and phonon energies is plotted in Fig. 3.4. We can find that the low frequency phonon couplings are generally much larger than high frequency phonons. Only several intramolecular phonons seem to be important. Their frequencies are in good agreement with Kato and Yamabe's calculation on single naphthalene molecules [18]. It is found that there exits only 13 modes with effective coupling constants larger than 0.01 [9]. Therefore, we list their detailed information in Table 3.2. From Table 3.3, we can see that for nonlocal electron–phonon coupling, the intermolecular vibration is much more important than the intramolecular modes, while for local electron–phonon coupling, the case is exactly the opposite.

3.2.2.3 Temperature Dependence of Mobility

Here we fix the lattice constant, and examine the temperature dependence of mobility for both electron and hole in a, b, and c' directions (see Fig. 3.5), where c' is perpendicular to the ab plane of naphthalene crystal. Hole is found to transport more efficiently than electron in naphthalene. The calculated temperature dependence agrees well with Karl's experiment except for electron in a and b directions [9, 19].

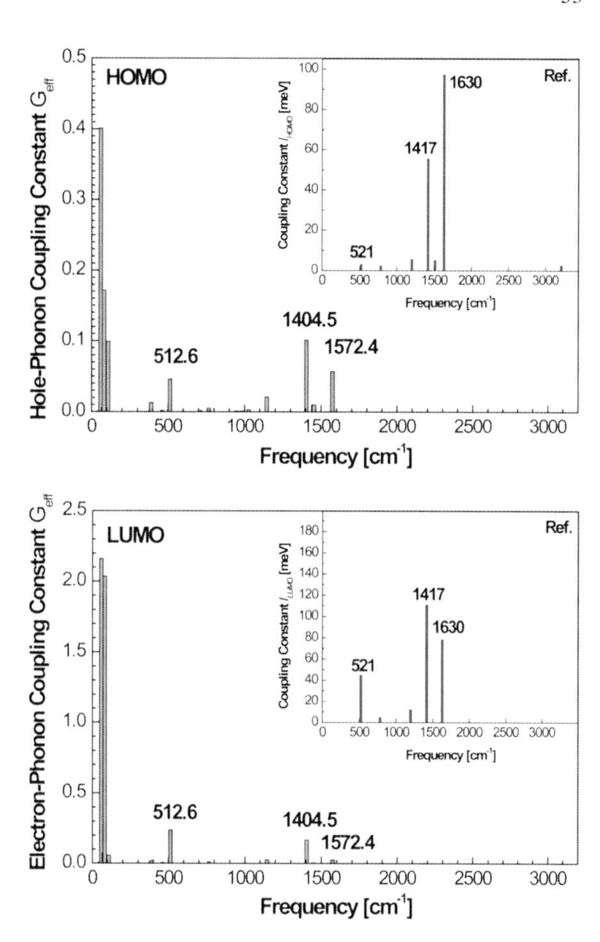

Fig. 3.4 Effective coupling constants vs phonon energies for electron (*up*) and hole (*down*) transport. Previous calculations for single naphthalene molecule by Kato and Yamabe [18] are shown in the insets. Reprinted with permission from Ref. [9]. Copyright 2007, American Institute of Physics

This is due to overestimated electron–phonon coupling constants related to the difficulties of DFT in describing weak van der Waals interaction. In Figs. 3.6 and 3.7, we show a detailed comparison between the calculated results and the experimental data for hole transport in all directions and electron transport in c' direction, respectively. We can find that hole mobilities generally decrease with temperature showing a band-like mechanism. Our calculated temperature dependence is in good agreement with the experiment below 60 K; however, at higher temperatures, the theoretical results show a slower decrease. For electron transport, we can clearly see a band-hopping transition, agreeing with the experiment. However, we notice that the calculated transition point is about 23 K, which is much lower than the experimentally measured 100–150 K range. In Sect. 3.3, we will see that the shortcomings can be overcome by using lattice constant changing with temperature.

Table 3.2 Calculated electron–phonon coupling constants for the most important 13 phonons through the first-principles-mapped tight-binding model

		0	*a*	*b*	*c*	*ac*	*ab*	*abc*
HOMO	g_{1mn}	−0.27	−0.09	0.49	0.13	0.13	−0.12	0.08
	g_{2mn}	0.15	−0.29	−0.13	0.19	0.05	−0.02	0.08
	g_{3mn}	−0.08	0.03	−0.12	0.03	0.08	−0.19	0.03
	g_{4mn}	0.00	0.00	0.01	0.00	0.00	−0.01	0.00
	g_{5mn}	0.05	0.02	−0.06	−0.02	−0.03	0.05	−0.03
	g_{6mn}	−0.05	0.00	−0.02	0.00	0.00	−0.01	0.00
	g_{7mn}	−0.21	−0.01	0.00	0.01	0.00	0.01	0.00
	g_{8mn}	0.06	0.02	0.03	0.00	0.00	−0.02	0.00
	g_{9mn}	−0.15	−0.01	−0.01	0.00	0.00	0.00	0.00
	g_{10mn}	0.32	0.00	0.00	0.00	0.00	0.00	0.00
	g_{11mn}	−0.10	0.00	0.00	0.00	0.00	0.01	0.00
	g_{12mn}	0.10	0.00	0.01	0.00	0.00	0.00	0.00
	g_{13mn}	0.24	0.00	−0.01	0.00	0.00	0.00	0.00
LUMO	g_{1mn}	−0.23	−0.04	−1.31	−0.31	0.01	0.39	0.00
	g_{2mn}	0.19	−0.04	0.20	0.66	−0.11	−0.86	0.10
	g_{3mn}	−0.09	−0.03	0.03	0.13	0.04	0.11	0.03
	g_{4mn}	0.10	0.01	−0.01	0.01	−0.01	−0.03	0.00
	g_{5mn}	−0.10	−0.01	0.04	−0.01	0.01	0.04	0.00
	g_{6mn}	0.15	0.01	0.01	−0.01	0.00	0.00	0.00
	g_{7mn}	0.49	0.00	0.00	0.01	0.00	0.01	0.00
	g_{8mn}	0.10	0.00	−0.02	0.00	0.00	0.00	0.00
	g_{9mn}	0.16	0.00	0.01	0.00	0.00	0.00	0.00
	g_{10mn}	−0.41	0.00	0.00	0.00	0.00	0.00	0.00
	g_{11mn}	0.06	0.00	0.00	0.00	0.00	0.00	0.00
	g_{12mn}	−0.05	0.00	0.00	0.00	0.00	0.00	0.00
	g_{13mn}	−0.16	0.00	0.01	0.00	0.00	−0.01	0.00

Table 3.3 The decompositions of the total effective coupling constant for HOMO and LUMO: local-inter, nonlocal-inter, local-intra, and nonlocal-intra parts

	HOMO	LUMO
$G_{Inter\text{-}Local}$	0.103	0.095
$G_{Inter\text{-}Nonlocal}$	0.569	4.158
$G_{Intra\text{-}Local}$	0.260	0.529
$G_{Intra\text{-}Nonlocal}$	0.024	0.101

3.2.2.4 Origin of Temperature Dependence of Mobility

As discussed above, the calculated temperature dependence of mobility is quite different for hole and electron in naphthalene. To understand the origin of the temperature dependence of mobility, we refer to the decomposition of mobility into a linear combination of various temperature- dependent functions, as proposed in Sect. 3.1.6. The temperature behavior of $f(T)$ and $h_q(T)$ for electron and hole transport is shown in Fig. 3.8. We can find that $f(T)$ generally decreases with temperature. For hole, the decrease is monotone, while for electron, after a sharp

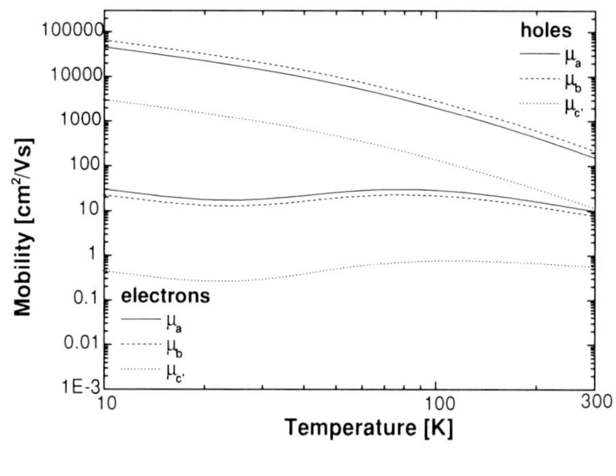

Fig. 3.5 Calculated charge-carrier mobilities as a function of temperature in naphthalene crystal from 10 to 300 K. The top three curves are for holes and the rest are for electrons. Reprinted with permission from Ref. [9]. Copyright 2007, American Institute of Physics

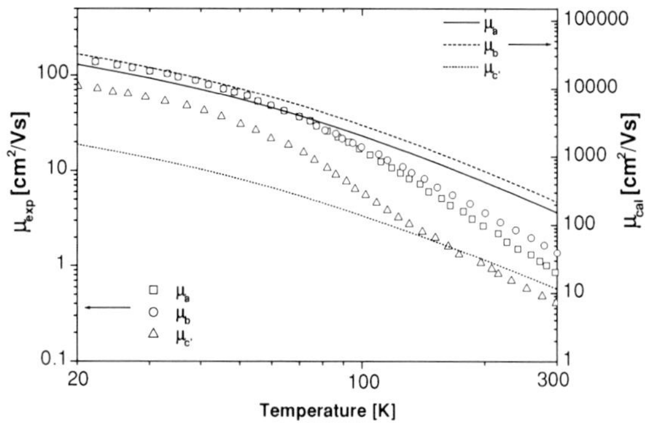

Fig. 3.6 The hole mobilities versus temperature obtained by calculation and experiments [19, 27] in *a*, *b*, and c' directions. Reprinted with permission from Ref. [9]. Copyright 2007, American Institute of Physics

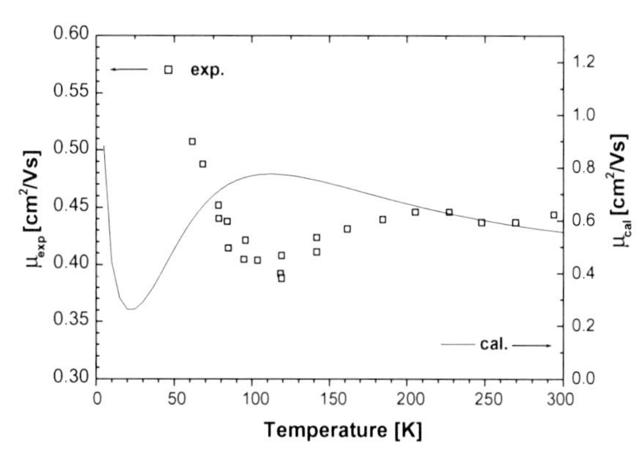

Fig. 3.7 Band-hopping transition for electron transport in the c' direction (theory vs. experiment [10]). Reprinted with permission from Ref. [9]. Copyright 2007, American Institute of Physics

decrease at low temperatures, it increases a little bit and decreases again at high temperatures. This behavior is because that the electron–phonon coupling strength for electron is much stronger than hole, as found in Table 3.3 [9]. For $h_q(T)$, it increases first with temperature and then levels off, and finally decreases with temperature, which is a typical character of classical Marcus hopping model as shown in Chap. 2. Besides, it is found that $h_q(T)$ decreases rapidly with phonons

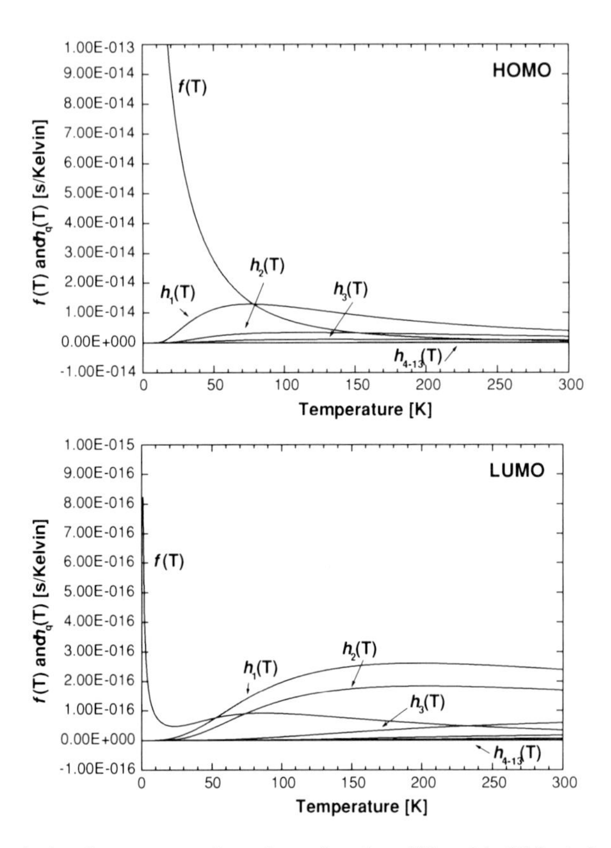

Fig. 3.8 The calculated temperature dependence functions *f(T)* and $h_q(T)$ for hole (HOMO) and electron (LUMO). Reprinted with permission from Ref. [9]. Copyright 2007, American Institute of Physics

frequencies, and thus only the lowest three modes from the intermolecular vibrations are important in this part. From the different temperature dependence behavior of *f(T)* and $h_q(T)$, we can phenomenologically tell that the band-like contribution comes from *f(T)* while the hopping contribution comes from $h_q(T)$. Therefore the relative magnitude of coefficients A and B_q determines the overall temperature dependence of mobility. From Table 3.4, we can see that B_q ranges from 10^{12} to 10^{14} (in cm^2K/Vs^2), which is much smaller than the range of A with 10^{16}–10^{17} cm^2K/Vs^2 for all directions. Therefore, the hole mobilities are mainly determined by the band-like function *f(T)*. For electron transport, we can find that in a and b directions, $A \gg B_q$, however in c', A is close in value to some of the B_q. This explains why band-hopping crossover can be seen for electron transport in the c' direction of naphthalene crystal. From the definition of A and B_q, we can conclude the band-like transport happens when the transfer integral is large and the nonlocal electron–phonon coupling is relatively small, otherwise, band-hopping crossover can be observed at a certain temperature range.

Table 3.4 The calculated coefficients A and B_q for hole (HOMO) and electron (LUMO) in a, b and c' directions (the unit is cm^2K/Vs^2)

	HOMO			LUMO		
	a	b	c'	a	b	c'
A	2.66E17	3.80E17	1.78E16	4.53E17	3.37E17	6.62E15
B_1	1.11E14	6.44E14	1.69E14	5.14E14	4.53E15	3.40E14
B_2	8.63E14	9.41E13	3.64E14	4.85E15	2.05E15	3.28E15
B_3	3.08E14	2.68E14	1.02E14	2.28E14	6.62E13	2.62E14
B_4	8.40E12	1.83E13	5.13E12	1.08E14	6.27E13	1.32E13
B_5	3.73E14	5.23E14	3.76E14	2.36E14	2.92E14	3.87E13
B_6	8.80E12	5.78E13	2.97E12	3.60E13	5.64E12	1.38E13
B_7	4.85E13	4.12E12	2.76E13	1.93E13	1.09E13	8.55E12
B_8	5.46E14	4.90E14	1.23E13	1.26E13	1.58E14	2.04E13
B_9	6.13E13	3.16E13	6.89E12	4.67E13	2.59E13	4.41E13
B_{10}	4.41E12	2.16E12	2.07E12	2.63E13	7.20E12	6.21E12
B_{11}	8.38E13	3.99E13	1.54E13	1.61E13	2.07E13	2.21E12
B_{12}	3.74E13	4.24E13	2.22E13	2.62E13	1.47E13	6.66E12
B_{13}	2.47E13	1.55E14	3.89E13	1.59E14	3.82E14	6.49E13

3.2.2.5 Temperature Dependence of Mobility

From the mobility formula in Eq. 3.26, we can find that the contribution from each phonon mode is determined by its electron–phonon coupling constant and phonon occupation number. As seen from Table 3.3, the major contributions to the total electron–phonon coupling constant come from intermolecular vibrations. Besides, low frequency modes have much more population over high frequency modes. Therefore, the intermolecular low frequency modes should determine the overall temperature dependence of mobility. To prove this, we compare the temperature dependence of mobility considering all phonons and that neglecting the intramolecular contributions (see Fig. 3.9). A reduction of about 69–80% mobility for electrons and 42–48% for holes has been observed from 10 to 300 K [9]. This reduction is almost temperature independent, supporting that the intermolecular vibrations are main contributions to the temperature dependence of mobility. The reduced amounts for electrons are more than for holes since the electron–phonon coupling constants of electrons are larger than those of holes.

3.3 Temperature Dependence of Mobility Considering Thermal Expansion

In principle, organic molecular systems are soft materials. With the increase of temperature, there is significant amount of lattice expansion, therefore the electronic properties and the charge mobility should change accordingly [20]. Here we numerically investigate how much the temperature dependence of mobility can be influenced by natural thermal expansion of lattice.

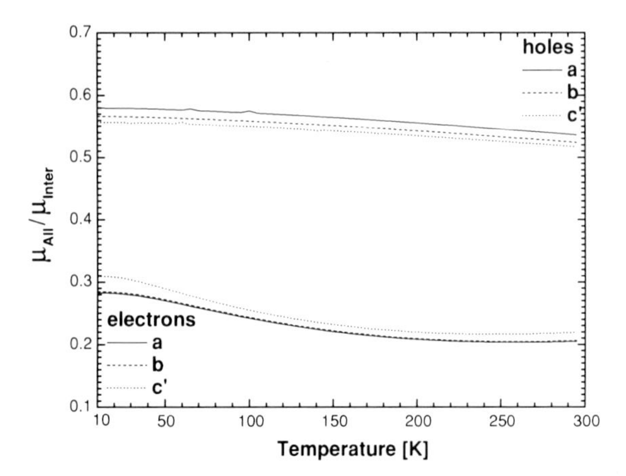

Fig. 3.9 The ratio between the calculated mobilities with all the phonons and the mobilities with only the intermolecular vibrations as function of temperature. The *top three curves* are for holes and the rest are for electrons. Reprinted with permission from Ref. [9]. Copyright 2007, American Institute of Physics

Table 3.5 Lattice parameters of naphthalene crystal at different temperatures (atmospheric pressures) ($\alpha = \gamma = 90°$)

Temperature [K]	a [Å]	b [Å]	c [Å]	β [°]
5	8.0711	5.9272	8.624	124.661
50	8.0798	5.9303	8.6288	124.582
92	8.1080	5.9397	8.6472	124.379
109	8.1224	5.9430	8.6525	124.322
120	8.1279	5.9461	8.6546	124.258
131	8.1356	5.9486	8.6568	124.197
143	8.1433	5.9512	8.6594	124.128
153	8.1508	5.9536	8.6610	124.066
163	8.1577	5.9559	8.6627	124.001
173	8.1647	5.9582	8.6644	123.933
184	8.1686	5.9617	8.6654	123.860
195	8.1799	5.9632	8.6678	123.772
206	8.1875	5.9657	8.6695	123.684
217	8.1952	5.9682	8.6711	123.591
228	8.2028	5.9707	8.6727	123.493
239	8.2128	5.9727	8.6745	123.388
273	8.2425	5.9806	8.6814	123.042
296	8.2606	5.9872	8.6816	122.671

3.3.1 Thermal Expansion of Lattice Constants

The lattice constants at different temperature are listed in Table 3.5. The experimental data of naphthalene crystal structures are available at temperatures $T = 5$, 50, 92, 109, 143, 184, 239, 273, 296 K [21, 22], where the data for $T = 5$ and 50 K is taken from the deuterated naphthalene, since it is found that its unit cell volume

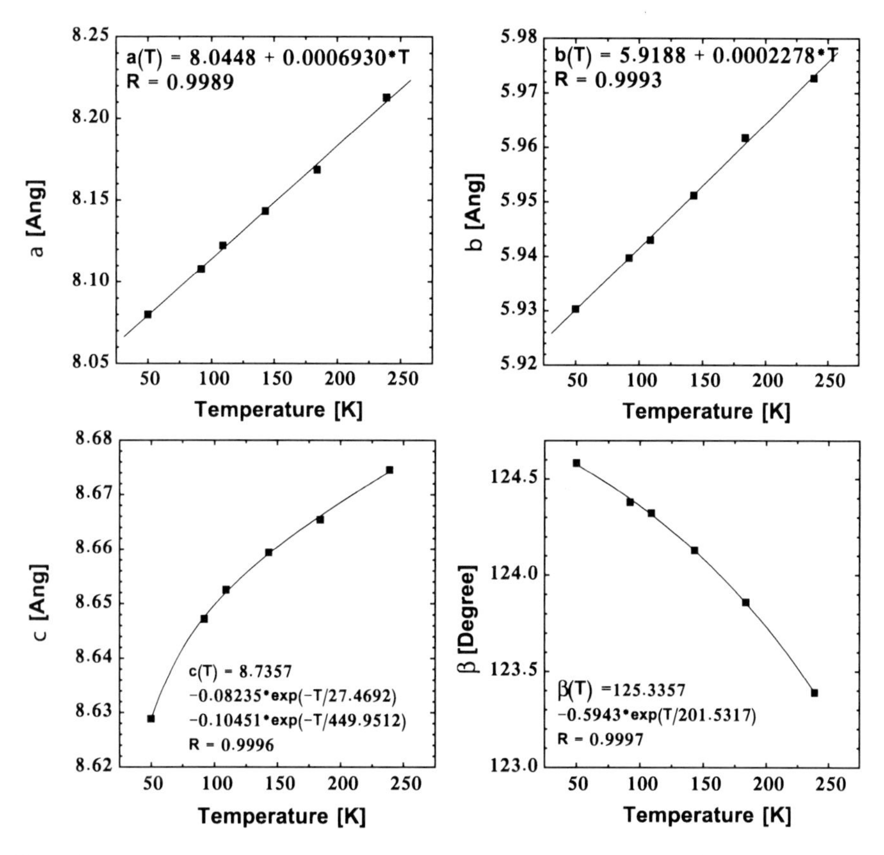

Fig. 3.10 Interpolating fits of crystal parameters $a(T)$, $b(T)$, $c(T)$ and $\beta(T)$ with available experimental data [21, 22] over the range 50–239 K. Reprinted with permission from Ref. [20]. Copyright 2008, American Institute of Physics

is only about 0.3–1.3% smaller than $C_{10}H_8$ due to the deuteration effect [22]. To get a smooth temperature dependence of lattice constant, we make appropriate interpolation among the experimental lattice constants for the whole temperature range by using analytic functions (see Fig. 3.10) [20]. It should be noted that due to limited experimental data, such interpolation does not reveal the physical law.

3.3.2 Temperature Dependence of Mobility with Structure Factor

With the temperature-dependent lattice constants, mobility at each temperature is recalculated with the same method as in Sect. 3.2. Here, we focus only on the hole transport in the b axis and electron transport in c' axis because the former has the largest mobility and the latter has a band-hopping transition behavior as shown

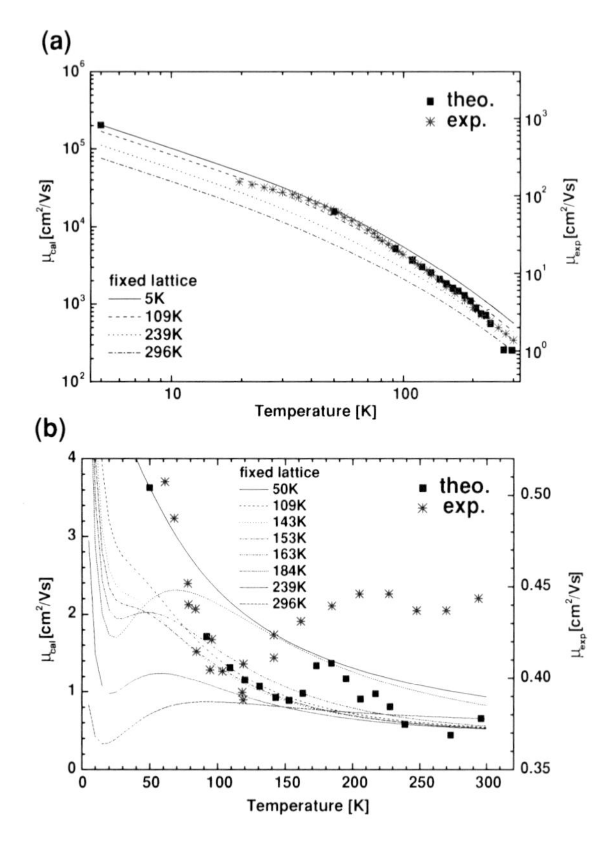

Fig. 3.11 The calculated and experimental [10, 19, 27] results of the temperature-dependent mobilities of **a** hole along b axis and **b** electron along c'-axis. The temperature-dependent mobilities with fixed lattice parameters are shown for comparison. Reprinted with permission from Ref. [20]. Copyright 2008, American Institute of Physics

Table 3.6 Lattice parameters of naphthalene crystal at different pressures (room temperature)

Pressure [GP]	a [Å]	b [Å]	c [Å]	β [°]
1.01325E−4	8.235	6.003	8.658	122.92
0.4	8.0348	5.8899	8.565	123.59
0.6	7.9948	5.8726	8.542	123.677
1.0	7.8523	5.8106	8.474	124.027
2.1	7.6778	5.721	8.395	124.55

previously. Calculated and experimental results are shown in Fig. 3.11. From hole transport, we can find that better agreement with experimental temperature dependence is achieved even within high temperature range. For electron transport, the band-hopping crossover temperature is calculated as 153 K [20], which is close to the experimental result of between 100 and 150 K [10]. Therefore we point out that the thermal expansion of lattice for organic molecular systems, which is generally neglected in theoretical studies for charge transport studies, can be very important to get the correct temperature dependence of mobility.

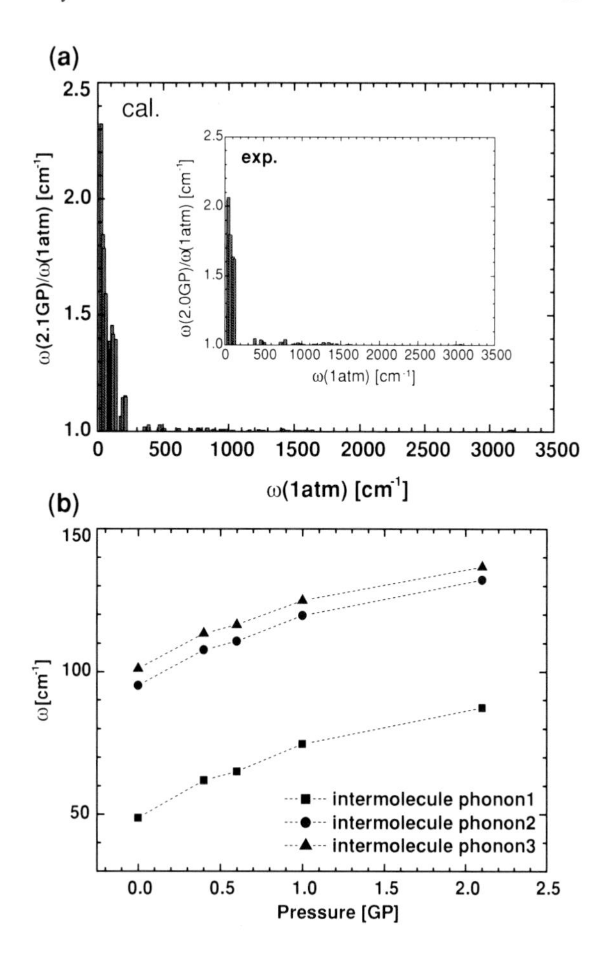

Fig. 3.12 **a** Ratio of the calculated phonon frequencies between 2.1 GPa with 1 atm. The experimental results [28] are shown in the inset. **b** Pressure dependence of the three intermolecular phonon frequencies. Reprinted with permission from Ref. [20]. Copyright 2008, American Institute of Physics

3.4 Pressure Dependence of Mobility

The lattice can be also strongly distorted at high pressures. In this section, we systematically investigate the role of pressure on electronic properties and charge mobility in naphthalene crystal.

3.4.1 Lattice Compression Under Pressure

Lattice constants under different pressure are listed in Table 3.6 [23, 24]. When the pressure is increased from the atmospheric pressure to 2.1 GPa, the reduction of lattice constant in *a*, *b*, and *c* axis are about 0.56, 0.28, and 0.26 Å, respectively, which is about three times larger than the effect of lowing temperature from 296 to 5 K as seen in Table 3.5. Therefore, the effect of pressure should be much stronger than that of temperature.

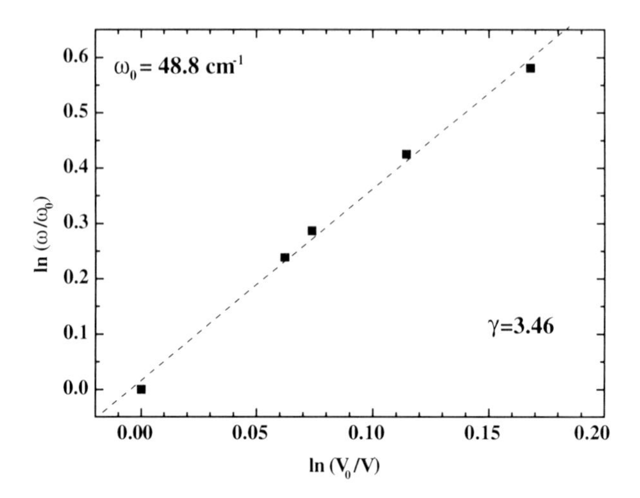

Fig. 3.13 Mode-Grüneisen parameter fit for the relationship between the phonon frequency and the unit-cell volume. Reprinted with permission from Ref. [20]. Copyright 2008, American Institute of Physics

3.4.2 Phonon Frequency

Under pressure, the lattice is more tightly packed, and thus the phonon frequencies should be increased. In Fig. 3.12a, we show the calculated and experimental pressure-induced phonon frequency change ratio between 2.1 GPa and 1 atm. A more explicit pressure dependence of the three intermolecular vibrational modes is shown in Fig. 3.12b. We can find that the most important changes occur at the low frequency part, which indicates that the pressure makes the crystal more solid and the intermolecular interaction is stronger.

Experimentally, the frequency change is usually related to the unit-cell volume change according to the mode-Grüneisen parameter:

$$\gamma = \frac{-d\left(\ln\frac{\omega}{\omega_0}\right)}{d\left(\ln\frac{V}{V_0}\right)} = \frac{d\left(\ln\frac{\omega}{\omega_0}\right)}{d\left(\ln\frac{V_0}{V}\right)} \tag{3.34}$$

For the most important intermolecular vibrational mode with frequency 48.8 cm^{-1}, we make a log–log plot to fit the constant γ, see Fig. 3.13. We get $\gamma = 3.46$ [20], which is in good agreement with the measured value $\gamma = 3.6$ [25].

3.4.3 Transfer Integral

As shown in Fig. 3.14, both the interlayer and intralayer transfer integrals are largely increased under pressure. Generally, the transfer integral is doubled when the pressure is increased from 1 atm to 2.1 GPa [20]. For some of the interlayer

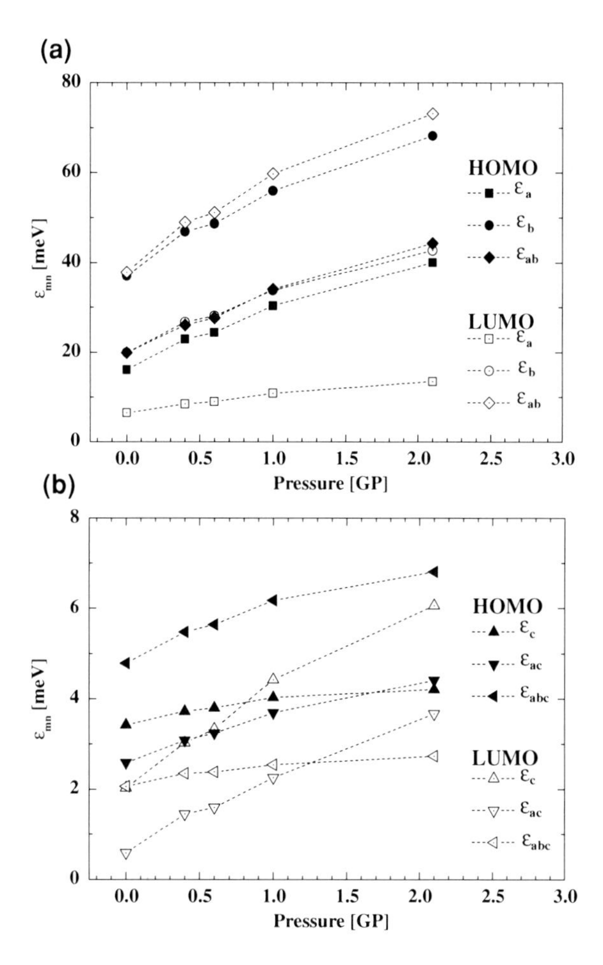

Fig. 3.14 Pressure-dependent transfer integrals of the **a** intralayer and **b** interlayer nearest neighbors. Reprinted with permission from Ref. [20]. Copyright 2008, American Institute of Physics

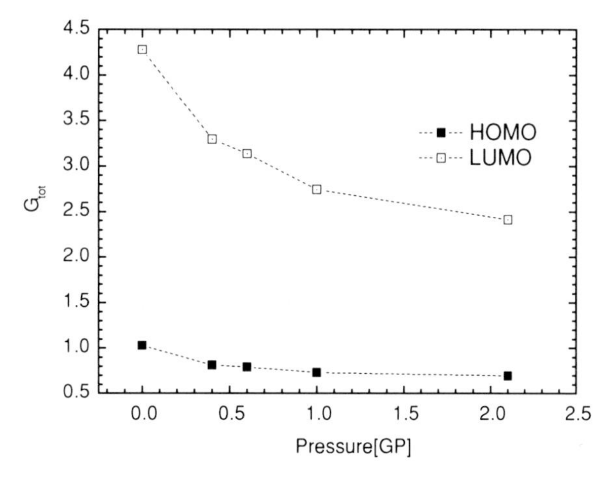

Fig. 3.15 Pressure dependence of the total electron–phonon coupling constants for hole (HOMO) and electron (LUMO). Reprinted with permission from Ref. [20]. Copyright 2008, American Institute of Physics

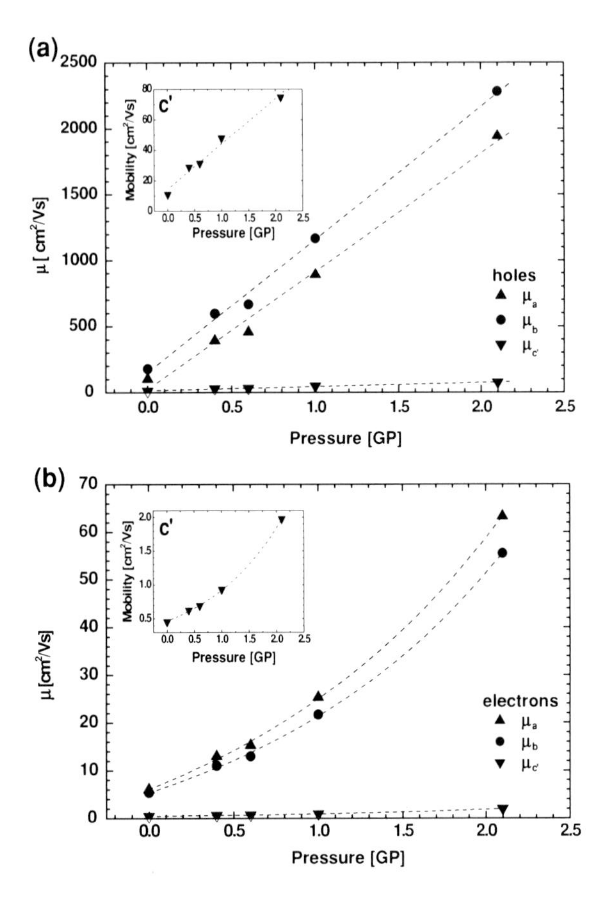

Fig. 3.16 Pressure dependence of the hole and the electron mobility. Reprinted with permission from Ref. [20]. Copyright 2008, American Institute of Physics

pathways, the increase of transfer integral can be even larger due to the weaker interaction between layers and thus high sensitivity of pressure.

3.4.4 Electron–Phonon Coupling

The pressure dependence of the total electron–phonon coupling constant for both electron and hole is plotted in Fig. 3.15. In both cases, the electron–phonon coupling decreases strongly about 40% when the pressure increases from 1 atm to 2.1 GPa. Since electron–phonon coupling constant for electron is much larger than that of hole in naphthalene, the former decreases in a more remarkable way.

3.4.5 Pressure Dependence of Mobility

We finally calculate the pressure-dependent mobilities for hole and electron along a, b, c' directions (see Fig. 3.16). It is found that the mobility increases linearly with pressure for holes, which is in good agreement with experiment [26]. For electron, mobility increases faster than hole. This should be related to the more significant drop of the electron–phonon coupling constant for electron than hole, and close behavior of the transfer integral upon pressure. From Fig. 3.16, we can see that the mobility is generally increased about one order of magnitude when a pressure of 2.1 GPa is applied. This again tells us that organic molecular crystals are very soft materials, and their charge transport properties are flexible with environmental change.

3.5 Conclusions

In this chapter, we have discussed the charge transport mechanism within the Holstein-Peierls polaron model. All kinds of electron–phonon coupling constants, including both local and nonlocal contributions from inter- and intra-molecular vibrations, have been taken into consideration through first-principles density functional theory calculations. For each of them, the electron–phonon coupling strength has been obtained, and their role on the temperature dependence of mobility has been systematically studied. We further examine the softness of molecular crystals, and its role on the reproduction of experimental temperature dependence of mobility. The effect of pressure on electronic properties and charge mobility has also been investigated. Overall, we find that (i) The low frequency intermolecular vibrations contribute most to the nonlocal electron–phonon couplings, while the high frequency intramolecular vibrations is important for local electron–phonon couplings; (ii) When transfer integral is large and nonlocal electron–phonon couplings are relatively weak, the charge transport is band-like, otherwise, the band-hopping crossover behavior can be observed; (iii) The intramolecular vibrations generally do not contribute to the overall temperature dependence of mobility; (iv) Thermal expansion of lattice is very important to get the correct temperature dependence of mobility agreeing with the experiment; (v) Mobility increases strongly with pressure. There are several points that we need to point out concerning the characteristics of the intermolecular vibrations which are of central importance. Firstly, these modes are generally anharmonic; however, all vibrations are considered as harmonic oscillators in the present studies. Secondly, the interaction between electron and phonon is assumed to be linear. Theoretical description including higher order interactions is also needed. Finally, we only consider Γ-point here, which is not enough to characterize the phonon dispersion of the low frequency acoustic phonons. In the next chapter, we will discuss another theory which is designed especially to study the acoustic phonon contributions to charge transport.

References

1. T. Holstein, Ann. Phys. (N.Y.) **8**, 343 (1959)
2. V. Coropceanu, J. Cornil, D.A. da Silva Filho, Y. Olivier, R. Silbey, J.-L. Brédas, Chem. Rev. **107**, 926 (2007)
3. R.W. Munn, R. Silbey, J. Chem. Phys. **83**, 1854 (1985)
4. V.M. Kenkre, J.D. Anderson, D.H. Dunlap, C.B. Duck, Phys. Rev. Lett. **62**, 1165 (1989)
5. K. Hannewald, P.A. Bobbert, Appl. Phys. Lett. **85**, 1535 (2004)
6. K. Hannewald, V.M. Stojanović, J.M.T. Schellekens, P.A. Bobbert, G. Kresse, J. Hafner, Phys. Rev. B **69**, 075211 (2004)
7. K. Hannewald, P.A. Bobbert, Phys. Rev. B **69**, 075212 (2004)
8. G.D. Mahan, *Many-Particle Physics* (Plenum Press, London, 1990)
9. L.J. Wang, Q. Peng, Q.K. Li, Z. Shuai, J. Chem. Phys. **127**, 044506 (2007)
10. L.B. Schein, C.B. Duck, A.R. McGhie, Phys. Rev. Lett. **40**, 197 (1978)
11. G. Kress, J. Hafner, Phys. Rev. B **47**, 558 (1993)
12. G. Kress, J. Hafner, Phys. Rev. B **49**, 14251 (1994)
13. G. Kress, J. Furthmüller, Phys. Rev. B **54**, 11169 (1996)
14. V.I. Ponomarev, O.S. Filipenko, L.O. Atovmyan, Crystallogr. Rep. **21**, 392 (1976)
15. J.P. Perdew, K. Burke, M. Ernzerho, Phys. Rev. Lett. **77**, 3865 (1996)
16. E.F.C. Byrd, G.E. Scuseria, C.F. Chabalowski, J. Phys. Chem. B **108**, 13100 (2004)
17. Y.C. Cheng, R.J. Silbey, D.A. da Silva Filho, J.P. Calbert, J. Cornil, J.-L. Brédas, J. Chem. Phys. **118**, 3764 (2003)
18. T. Kato, T. Yamabe, J. Chem. Phys. **115**, 8592 (2001)
19. N. Karl, in *Organic Semiconductors, Landolt Bornstein, New Series* Group III, vol. 17, ed. by K.-H. Hellwege, O. Madelung (Springer, Berlin, 1985), pp. 106–218
20. L.J. Wang, Q.K. Li, Z. Shuai, J. Chem. Phys. **128**, 194706 (2008)
21. C.P. Brock, J.D. Dunitz, Acta Crystallogr. Sect. B Struct. Crystalogr. Cryst. Chem. **38**, 2218 (1982)
22. E. Baharie, G.S. Pawley, Acta Crystallogr. Sect. A Cryst. Phys. Diffr. Theor. Gen. Crystallogr. **38**, 803 (1982)
23. F.R. Ahmed, D.W.J. Cruickshank, Acta Crystallogr. **5**, 852 (1952)
24. F.P.A. Fabbiani, D.R. Allan, S. Parsons, C.R. Pulham, Acta Crystallogr. Sect. B Struct. Sci. **62**, 826 (2006)
25. W. Häfner, W. Kiefer, J. Chem. Phys. **86**, 4582 (1987)
26. Z. Rang, M.I. Nathan, P.P. Ruden, V. Podzorov, M.E. Gershenson, C.R. Newman, C.D. Frisbie, Appl. Phys. Lett. **86**, 123501 (2005)
27. W. Warta, N. Karl, Phys. Rev. B **32**, 1172 (1985)
28. M. Nicol, M. Vernon, J.T. Woo, J. Chem. Phys. **63**, 1992 (1975)

Chapter 4
Deformation Potential Theory

Abstract When the electron–phonon coupling is weak compared with the intermolecular electronic couplings, charge transport can be described by the band mechanism. Namely, the charge moves coherently in a wavelike manner and is scattered by phonon. In this chapter, we introduce the deformation potential theory, which is actually a band model including only the lattice scatterings by the acoustic deformation potential. It is based on the Boltzmann transport equation and sometimes, can be simplified using the effective mass approximation. Contrary to Chap. 3, where only optical phonons are considered, the acoustic phonons are the focus of this chapter. This approach is applied to a typical molecular crystal, naphthalene, and covalently bonded functional materials, graphene and graphdiyne sheets and nanoribbons.

Keywords Deformation potential theory · Band model · Boltzmann transport equation · Effective mass approximation · Acoustic phonon · Graphene and graphdiyne sheets and nanoribbons

In Sect. 4.1, we derive the mobility formula of the deformation potential theory using the Boltzmann transport equation and the effective mass approximation and discuss how the parameters in the mobility expression are calculated first principally. In Sect. 4.2, the above method is applied to oligoacenes to study the role of acoustic phonons, acting as a complement to the studies in previous chapters. Applications to graphene and graphdiyne are discussed in Sect. 4.3 and Sect. 4.4, respectively.

Z. Shuai et al., *Theory of Charge Transport in Carbon Electronic Materials*,
SpringerBriefs in Molecular Science, DOI: 10.1007/978-3-642-25076-7_4,

4.1 Deformation Potential Theory

The deformation potential theory was proposed by Bardeen and Shockley about 60 years ago to investigate the role of acoustic phonons on electron and hole mobilities in nonpolar inorganic semiconductors like silicon, germanium and tellurium [1]. The basic argument is that for single crystal silicon, or other inorganic semiconductor, the electron wave is coherent with a thermal wavelength much longer than the lattice constant. Thus the primary scattering comes from the long wavelength acoustic phonons, and the scattering can be approximated by a uniform lattice dilation, or uniform deformation. Here we briefly describe the basic ideas of this approach for studying charge transport properties in organic materials.

4.1.1 Standard Form of Boltzmann Transport Equation

The Boltzmann transport equation is a general tool to analyze transport phenomena. It describes the time evolution of the distribution function $f(\boldsymbol{r},\boldsymbol{k},t)$ for one particle in the phase space, where $\boldsymbol{r}$ and $\boldsymbol{k}$ are position and momentum of the particle, respectively. Generally, the total time derivative of $f(\boldsymbol{r},\boldsymbol{k},t)$ can be expressed as

$$\frac{df}{dt} = \frac{\partial f}{\partial t} + \frac{\partial f}{\partial \boldsymbol{r}}\frac{d\boldsymbol{r}}{dt} + \frac{\partial f}{\partial \boldsymbol{k}}\frac{d\boldsymbol{k}}{dt} \tag{4.1}$$

Since $\mathrm{d}\boldsymbol{r}/\mathrm{d}t = \boldsymbol{v}$ is the velocity and $\hbar \mathrm{d}\boldsymbol{k}/\mathrm{d}t = \boldsymbol{F}$ is the external force acting on the particle, we have

$$\frac{df}{dt} = \frac{\partial f}{\partial t} + \frac{\partial f}{\partial \boldsymbol{r}}\boldsymbol{v}(\boldsymbol{k}) + \frac{\partial f}{\partial \boldsymbol{k}}\frac{\boldsymbol{F}(\boldsymbol{r})}{\hbar} \tag{4.2}$$

Here, we adopt a reasonable assumption that the velocity and the external force are only related to momentum and position of the particle, respectively. If the scattering between different electronic states is the only mechanism to balance the distribution function change due to electron diffusion and external force, we can set

$$\frac{df}{dt} = \left.\frac{\partial f}{\partial t}\right|_{scatt} \tag{4.3}$$

Combining Eq. 4.2 and Eq. 4.3, we can derive the standard form of the Boltzmann transport equation,

$$\left.\frac{\partial f}{\partial t}\right|_{scatt} = \frac{\partial f}{\partial t} + \frac{\partial f}{\partial \boldsymbol{r}}\boldsymbol{v}(\boldsymbol{k}) + \frac{\partial f}{\partial \boldsymbol{k}}\frac{\boldsymbol{F}(\boldsymbol{r})}{\hbar} \tag{4.4}$$

4.1.2 Boltzmann Transport Function for Charge Transport

For an electronic system, the distribution function at equilibrium adopts the Fermi–Dirac distribution,

$$f_0 = \frac{1}{\exp\{[\varepsilon(\boldsymbol{k}) - E_F]/k_B T\} + 1} \tag{4.5}$$

where $\varepsilon(\boldsymbol{k})$ describes the energy band for the charge carrier, namely, the valence band (VB) for hole and the conduction band (CB) for electron, and E_F is the Fermi level (also called chemical potential) of the system. At weak electric field limit, the distribution function f should be very close to f_0. Therefore we normally assume that f depends only on $\boldsymbol{k}$, the same as f_0, and then we can neglect the first and second terms on the right side of Eq. 4.4. For charge transport, $\boldsymbol{F}(\boldsymbol{r}) = -e_0\boldsymbol{E}$, where e_0 is the element charge of an electron and $\boldsymbol{E}$ is the electric field. Considering that $\boldsymbol{v}(\boldsymbol{k}) = 1/\hbar \times \partial\varepsilon(\boldsymbol{k})/\partial\boldsymbol{k}$ is the group velocity, Eq. 4.4 becomes

$$\left.\frac{\partial f}{\partial t}\right|_{scatt} = -\frac{\partial f}{\partial \boldsymbol{k}}\frac{e_0\boldsymbol{E}}{\hbar} = -\frac{\partial f}{\partial \varepsilon}\frac{\partial \varepsilon}{\partial \boldsymbol{k}}\frac{e_0\boldsymbol{E}}{\hbar} = -e_0\boldsymbol{E}\boldsymbol{v}(\boldsymbol{k})\frac{\partial f}{\partial \varepsilon} \approx -e_0\boldsymbol{E}\boldsymbol{v}(\boldsymbol{k})\frac{\partial f_0}{\partial \varepsilon} \tag{4.6}$$

4.1.3 Relaxation Time Approximation

Normally, the scattering term can be expressed as

$$\left.\frac{\partial f}{\partial t}\right|_{scatt} = \sum_{k'} \{\Theta(\boldsymbol{k}',\boldsymbol{k})f(\boldsymbol{k}')[1 - f(\boldsymbol{k})] - \Theta(\boldsymbol{k},\boldsymbol{k}')f(\boldsymbol{k})[1 - f(\boldsymbol{k}')]\} \tag{4.7}$$

where $\Theta(\boldsymbol{k},\boldsymbol{k}')$ is the transition probability from electronic state $\boldsymbol{k}$ to $\boldsymbol{k}'$. At thermal equilibrium, we have $\Theta(\boldsymbol{k}',\boldsymbol{k}) = \Theta(\boldsymbol{k},\boldsymbol{k}')$, and thus Eq. 4.7 can be simplified as

$$\left.\frac{\partial f}{\partial t}\right|_{scatt} = \sum_{k'} \Theta(\boldsymbol{k},\boldsymbol{k}')[f(\boldsymbol{k}') - f(\boldsymbol{k})] \tag{4.8}$$

When the relaxation time approximation is applied [2], the scattering term can be also expressed as

$$\left.\frac{\partial f}{\partial t}\right|_{scatt} = -\frac{f(\boldsymbol{k}) - f_0(\boldsymbol{k})}{\tau(\boldsymbol{k})} \tag{4.9}$$

Inserting Eq. 4.9 into Eq. 4.6, $f(\boldsymbol{k})$ can be expressed with $\tau(\boldsymbol{k})$ as

$$f(\boldsymbol{k}) \approx f_0(\boldsymbol{k}) + e_0\tau(\boldsymbol{k})\boldsymbol{E}\cdot\boldsymbol{v}(\boldsymbol{k})\frac{\partial f_0}{\partial \varepsilon} \tag{4.10}$$

From Eqs. 4.8 and 4.9, we have

$$-\frac{f(\boldsymbol{k})-f_0(\boldsymbol{k})}{\tau(\boldsymbol{k})}=\sum_{k'}\Theta(\boldsymbol{k},\boldsymbol{k}')[f(\boldsymbol{k}')-f(\boldsymbol{k})] \tag{4.11}$$

Thus

$$\frac{1}{\tau(\boldsymbol{k})}=\sum_{k'}\Theta(\boldsymbol{k},\boldsymbol{k}')\left[1+\frac{f_0(\boldsymbol{k})-f(\boldsymbol{k}')}{f(\boldsymbol{k})-f_0(\boldsymbol{k})}\right] \tag{4.12}$$

We assume that the scattering is elastic, which means that $\varepsilon(\boldsymbol{k})=\varepsilon(\boldsymbol{k}')$ and thus $f_0(\boldsymbol{k})=f_0(\boldsymbol{k}')$, and then inserting Eq. 4.10 into Eq. 4.12, we can get

$$\frac{1}{\tau(\boldsymbol{k})}\approx\sum_{k'}\Theta(\boldsymbol{k},\boldsymbol{k}')\left[1-\frac{\tau(\boldsymbol{k}')\boldsymbol{v}(\boldsymbol{k}')\cdot\boldsymbol{e}_E}{\tau(\boldsymbol{k})\boldsymbol{v}(\boldsymbol{k})\cdot\boldsymbol{e}_E}\right] \tag{4.13}$$

where $\boldsymbol{e}_E$ is the unit vector along the electric field. In principle, Eq. 4.13 can be solved iteratively. Here we normally further simplify Eq. 4.13 as

$$\frac{1}{\tau(\boldsymbol{k})}\approx\sum_{k'}\Theta(\boldsymbol{k},\boldsymbol{k}')\left[1-\frac{\boldsymbol{v}(\boldsymbol{k}')\cdot\boldsymbol{e}_E}{\boldsymbol{v}(\boldsymbol{k})\cdot\boldsymbol{e}_E}\right] \tag{4.14}$$

4.1.4 Deformation Potential

According to the Fermi's golden rule, the transmission probability can be expressed as

$$\Theta(\boldsymbol{k},\boldsymbol{k}')=\frac{2\pi}{\hbar}|M(\boldsymbol{k},\boldsymbol{k}')|^2\delta[\varepsilon(\boldsymbol{k})-\varepsilon(\boldsymbol{k}')] \tag{4.15}$$

Here $M(\boldsymbol{k},\boldsymbol{k}')=\langle \boldsymbol{k}|\Delta V(\boldsymbol{r})|\boldsymbol{k}'\rangle$, where $|\boldsymbol{k}\rangle$ is the Bloch wave function of the electron with wave vector $\boldsymbol{k}$ and $\Delta V(\boldsymbol{r})$ is the potential perturbation due to thermal motions. The deformation potential theory assumes that $\Delta V(\boldsymbol{r})$ has a linear dependence on the relative volume change, $\Delta(\boldsymbol{r})$, namely,

$$\Delta V(\boldsymbol{r})=E_1\Delta(\boldsymbol{r}) \tag{4.16}$$

where E_1 is named as the deformation potential constant. The acoustic phonon with wave vectors $\{\boldsymbol{q}\}$ and the displacement for lattice site $\boldsymbol{r}$ is

$$\boldsymbol{u}(\boldsymbol{r})=\frac{1}{\sqrt{N}}\sum_{q}\boldsymbol{e}_q\left[a_q e^{i\boldsymbol{qr}}+a_q^* e^{-i\boldsymbol{qr}}\right] \tag{4.17}$$

where N is the number of lattice sites in the unit volume. $\boldsymbol{e}_q$ and a_q are the unit vector and amplitude of the acoustic phonon $\boldsymbol{q}$, respectively. At high temperatures, when the lattice waves are fully excited, the amplitude of the wave is given by $|a_q|^2 = k_B T/2M\boldsymbol{q}^2 v_q^2$ according to the uniform energy partition theory [1], where M is the total mass of lattice in the unit volume, and v_q is the velocity of the wave. Then the relative volume change could be expressed as

$$\Delta(\boldsymbol{r}) \equiv \frac{\partial \boldsymbol{u}(\boldsymbol{r})}{\partial \boldsymbol{r}} = i\frac{1}{\sqrt{N}}\sum_q \boldsymbol{q}\cdot\boldsymbol{e}_q\left[a_q e^{iqr} - a_q^* e^{-iqr}\right] \tag{4.18}$$

With Eqs. 4.16 and 4.18, the electronic coupling element can be calculated as [1]

$$|M(\boldsymbol{k},\boldsymbol{k}')|^2 = N^{-1}E_1^2\boldsymbol{q}^2 a_q^2 = \frac{k_B T E_1^2}{c_q} \tag{4.19}$$

where $\boldsymbol{q} = \pm(\boldsymbol{k}' - \boldsymbol{k})$, $c_q = 2NMv_q^2$ is the elastic constant or the strength modulus. Then Eq. 4.15 becomes

$$\Theta(\boldsymbol{k},\boldsymbol{k}') = \frac{2\pi k_B T E_1^2}{\hbar c_q}\delta[\varepsilon(\boldsymbol{k}) - \varepsilon(\boldsymbol{k}')] \tag{4.20}$$

and correspondingly Eq. 4.14 can be expressed as [3]

$$\frac{1}{\tau(\boldsymbol{k})} \approx \frac{2\pi k_B T E_1^2}{\hbar c_q}\sum_{k'}\left[1 - \frac{\boldsymbol{v}(\boldsymbol{k}')\cdot\boldsymbol{e}_E}{\boldsymbol{v}(\boldsymbol{k})\cdot\boldsymbol{e}_E}\right]\delta[\varepsilon(\boldsymbol{k}) - \varepsilon(\boldsymbol{k}')] \tag{4.21}$$

4.1.5 Mobility Formula

By definition, the mobility (μ_α) is the ratio between the drift velocity, v_α, and the electric field, E_α,

$$\mu_\alpha = \frac{v_\alpha}{E_\alpha} = \frac{\int v_\alpha(\boldsymbol{k})f(\boldsymbol{k})d\boldsymbol{k}}{E_\alpha\int f(\boldsymbol{k})d\boldsymbol{k}} \tag{4.22}$$

where α is the electric field direction. Substituting Eq. 4.10 into Eq. 4.22 and considering that $f_0(\boldsymbol{k})$ is an even function, while $v(\boldsymbol{k})$ is an odd function, we have

$$\int v_\alpha(\boldsymbol{k})f(\boldsymbol{k})d\boldsymbol{k} = e_0 E_\alpha\int \tau(\boldsymbol{k})v_\alpha^2(\boldsymbol{k})\frac{\partial f_0}{\partial\varepsilon}d\boldsymbol{k} \tag{4.23}$$

If we consider all the energy bands for the charge carrier, Eq. 4.23 becomes

$$\int v_\alpha(\boldsymbol{k})f(\boldsymbol{k})d\boldsymbol{k} = e_0 E_\alpha\sum_{i\in CB(VB)}\int \tau(i,\boldsymbol{k})v_\alpha^2(i,\boldsymbol{k})\frac{\partial f_0}{\partial\varepsilon}d\boldsymbol{k} \tag{4.24}$$

for electron (hole), where $\tau(i,\boldsymbol{k})$ and $v_\alpha(i,\boldsymbol{k})$ are the relaxation time and the group velocity of the ith band with wave vector $\boldsymbol{k}$. Correspondingly, the integral on the denominator of Eq. 4.22 for electron and hole can be expressed as

$$\int f(\boldsymbol{k})d\boldsymbol{k} \approx \sum_{i\in CB} \int f_0[\varepsilon_i(\boldsymbol{k}) - E_F]d\boldsymbol{k} \tag{4.25}$$

and

$$\int f(\boldsymbol{k})d\boldsymbol{k} \approx \sum_{i\in VB} \int \{1 - f_0[\varepsilon_i(\boldsymbol{k}) - E_F]\}d\boldsymbol{k} \tag{4.26}$$

respectively. Normally the band gap is much larger than k_BT, and thus for $i \in CB$ $\exp\{[\varepsilon_i(\boldsymbol{k})\text{-}E_F]/k_BT\} >> 1$, we can replace the Fermi–Dirac distribution with the Boltzmann distribution, $f_0 \approx \exp\{-[\varepsilon_i(\boldsymbol{k}) - E_F]/k_BT\}$. Finally the electron (hole) mobility can be expressed as [3]

$$\mu_\alpha^{e(h)} = \frac{e_0}{k_BT} \frac{\sum\limits_{i\in CB(VB)} \int \tau(i,\boldsymbol{k}) v_\alpha^2(i,\boldsymbol{k}) \exp[\mp\varepsilon_i(\boldsymbol{k})/k_BT]d\boldsymbol{k}}{\sum\limits_{i\in CB(VB)} \int \exp[\mp\varepsilon_i(\boldsymbol{k})/k_BT]d\boldsymbol{k}} \tag{4.27}$$

4.1.6 Effective Mass Approximation

The effective mass approximation can be used to simplify the mobility formulas. In 1D systems, the energy band can be written in very simple form as

$$\varepsilon(k) = \varepsilon_0 + \left(\hbar^2/2m^*\right)(k - k_0)^2 \tag{4.28}$$

where m^* is the effective mass. Using $v(k) = \hbar k/m^*$, Eq. 4.21 becomes

$$\frac{1}{\tau(k)} = \frac{2E_1^2 k_BTm^*}{k\hbar^3 c_q} \tag{4.29}$$

Then the mobility can be calculated through [4]

$$\mu_{1D} = \frac{e_0 \int \tau(k)f(k)dk}{m^* \int f(k)dk} = \frac{e_0\hbar^2 c_q}{E_1^2(2\pi k_BT)^{1/2} m^{*3/2}} \tag{4.30}$$

Similarly, the mobility in the 2D and 3D cases can be derived as [1]

$$\mu_{2D} = \frac{2e_0\hbar^3 c_q}{3E_1^2 k_BTm^{*2}} \tag{4.31}$$

$$\mu_{3D} = \frac{2(2\pi)^{1/2} e_0 \hbar^4 c_q}{3E_1^2 (k_B T)^{3/2} m^{*5/2}} \tag{4.32}$$

In the 3D case, the temperature dependence of mobility in Eq. 4.32 follows a power law with factor −1.5, in close agreement with experiment [5, 6]. Note that the m^* in Eq. 4.32 is an average over all directions, which is not appropriate to describe the anisotropic behavior of mobility.

4.1.7 Numerical Parameterization

To obtain the mobility from Eq. 4.27, there are several parameters to be determined. The energy band $\varepsilon_i(\boldsymbol{k})$ can be calculated by the first principle density functional theory. By following the work of Madsen and Singh [7], the group velocities $v_\alpha(i,\boldsymbol{k})$ are obtained through numerical differentiation in the k-space with smoothed Fourier interpolation. The anisotropic relaxation time $\tau(i,\boldsymbol{k})$ can be evaluated with Eq. 4.21, with the known of the elastic constant, c_α and the deformation potential constant, E_1. By fitting of the curve of total energy change per volume, $\Delta E/V_0$ to dilation $\Delta l/l_0$ with formula $\Delta E/V_0 = c_\alpha(\Delta l/l_0)^2/2$, we can evaluate c_α along the transport direction α. Here V_0 is the cell volume at equilibrium, and l_0 is the lattice constant along α direction. A typical example of the fitting of c_α is shown in Fig. 4.1a. E_1 is defined as $E_1 = \Delta V_i/(\Delta l/l_0)$ where ΔV_i is energy change of the ith band with lattice dilation $\Delta l/l_0$ along the direction of external electric field. As shown in Fig. 4.1b, for the sake of simplicity, we generally take the energy change at conduction band minimum (CBM) and at valence band maximum (VBM) for electron and hole, respectively. Besides, following the approach proposed by Wei and Zunger [8], we assume that the localized 1s level is not sensitive to slight lattice deformation and can be used as a reference to obtain the absolute band energy changes for both VBM and CBM. In Eq. 4.30, we also need the effective mass m^*. It can be calculated through a quadratic fit in the energy versus k-points for the bottom (top) of CB (VB) for electron (hole).

4.2 Application: Oligoacenes

Several oligoacenes have been investigated in Chap. 2 with the hopping mechanism and in Chap. 3 with the polaron mechanism, both of which regard all nuclear vibrations as optic phonons. Hereby, we discuss the role of acoustic phonon scattering on the charge transport properties with deformation potential theory. The crystal structures for the investigated oligoacenes, i.e., naphthalene, anthracene, tetracene and pentacene, are shown in Fig. 4.2.

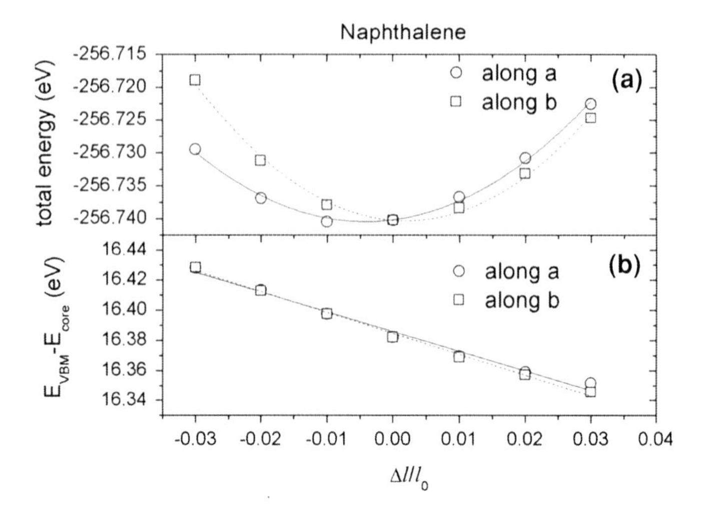

Fig. 4.1 A typical plot to obtain **a** the elastic constants and **b** the deformation potential constants through fitting **a** the total energy of unit cell with uniform dilation $\Delta l/l_0$ and **b** the band energy of VBM (E_{VBM}) with respect to the closest level to the core level (E_{core}) with uniform dilation. The lines are the fitting parabola. The plot is taken from calculations for naphthalene along a and b directions [3]. Reproduced from Ref. [3] with permission by Springer

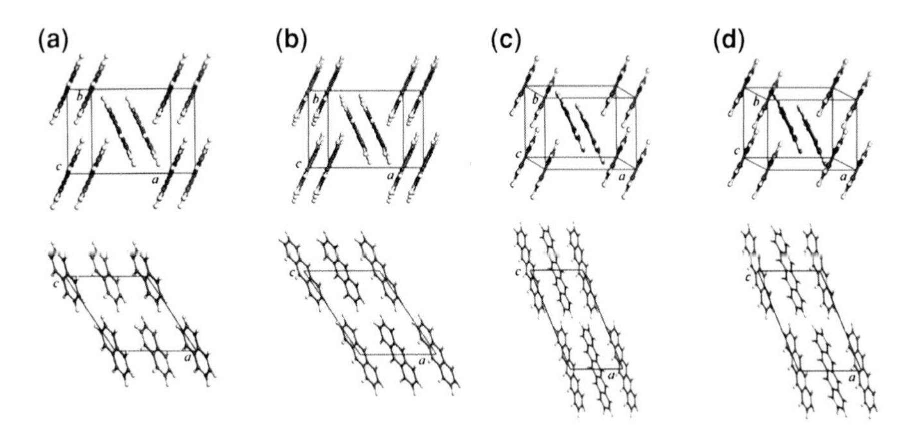

Fig. 4.2 Crystal structures of oligoacenes: **a** naphthalene, **b** anthracene, **c** tetracene and **d** pentacene. Reproduced from Ref. [3] with permission by Springer

4.2.1 Computational Details

The initial crystal structures of oligoacenes are taken from the Cambridge Structural Database [9–11]. The Vienna ab initio Simulation Package (VASP) [12–14] is used to optimize the lattice and to obtain the energy bands together with LDA exchange–correlation potential. The discrepancy in lattice constants between the

Table 4.1 The calculated elastic constants C [10^{10} dyn/cm^2] and deformation potential constants E [eV] of oligoacene crystals

	Naphthalene	Anthracene	Tetracene	Pentacene
C_a	16.24	16.36	13.80	14.45
C_b	21.10	19.93	19.86	19.83
E_a^h	1.31	1.12	1.79	2.10
E_b^h	1.39	1.38	0.47	0.79
E_a^e	0.96	0.42	1.60	1.81
E_b^e	0.56	0.87	0.53	0.38

LDA results and the experimental values is within 5% [3]. The mobility formula without effective mass approximation, Eq. 4.27, is adopted.

4.2.2 Results and Discussion

4.2.2.1 Elastic Constants and Deformation Potential Constants

The fitting procedure described in Sect. 4.1.7 is used to get the elastic constants and deformation potential constants for all the investigated oligoacene crystals (see Table 4.1). Only the results along a and b lattice axes are shown since it is believed that the in-plane mobilities are more useful to the device applications. Agreeing with the crystal symmetry, the properties of naphthalene and anthracene are generally close except the deformation potential constants for electron. Similarly, the values for tetracene and pentacene are also quite close.

4.2.2.2 Temperature Dependence of Mobility

The temperature dependent hole and electron mobilities in a and b directions are shown in Fig. 4.3. Overall, the mobility manifests the typical power law, owing to the intrinsic band transport mechanism of the deformation potential theory. The calculated temperature dependence can be approximated as $\mu \propto T^{-1.5}$ and the deviation is the effect beyond the effective mass approximation.

4.2.2.3 Electron and Hole Mobilities of Oligoacenes

In Table 4.2, we list the calculated electron and hole room temperature mobilities of all investigated oligoacene crystals. The mobility shows no apparent molecular length dependence due to their different crystal packings. In Chap. 3, we only consider optic phonon scattering processes and very small inhomogeneous external scattering, the room temperature hole mobilities in naphthalene are calculated to be about 150–200 cm^2/Vs [15]. In contrast, here we only take acoustic phonons into

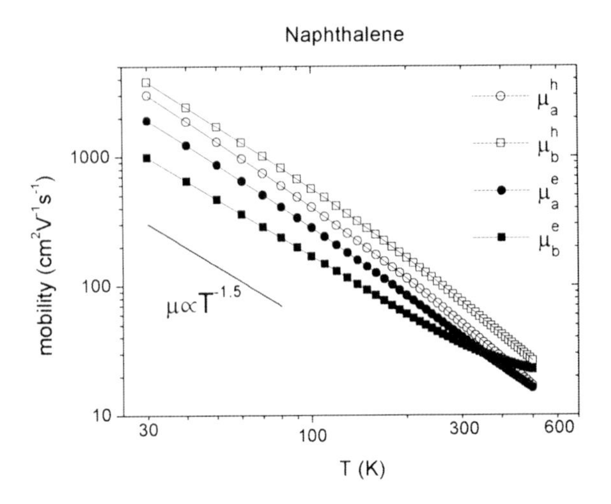

Fig. 4.3 Temperature dependence of hole and electron mobilities along a and b directions for naphthalene. The power law with factor −1.5 is shown in the figure which describes the temperature dependence of carrier with effective mass approximation. Reproduced from Ref. [3] with permission by Springer

Table 4.2 The calculated carrier mobility in unit cm^2/Vs at T = 300 K of oligoacene crystals

	Naphthalene	Anthracene	Tetracene	Pentacene
μ_a^h	50.4	19.2	10.6	15.2
μ_b^h	74.4	42.2	92.5	55.6
μ_a^e	39.8	245	24.5	27.7
μ_b^e	35.3	15.4	87.6	295

account, and the hole mobilities are around 50–75 cm^2/Vs, which are about three times smaller than that with only optic phonons. In other words, the acoustic phonon scattering mechanism is about three times as strong as that with the optic phonons in naphthalene, suggesting the important role of acoustic phonon in charge transport. Generally, organic semiconductors are of p-type in field effect transistors, which means that the measured hole mobility is larger than electron mobility. This is primarily due to the fact that (1) the electrode (e.g. gold) work function is closer to the highest occupied molecular orbitals so that the charge injection consists of mostly holes and (2) organic materials present much more electron traps than hole traps [3]. Hereby, we find that the calculated intrinsic electron mobility (e.g. anthracene in a direction and pentacene in b direction) can be even larger than that of hole, as shown in Table 4.2. This is mostly due to their much smaller electron deformation potential constants as shown in Table 4.1.

4.3 Application: Graphene

Graphene is a one-atom-thick planar sheet of sp^2-bonded carbon atoms which are densely packed in a two-dimensional honeycomb lattice (see Fig. 4.4a). Individual graphene sheets were first isolated by Novoselov and Geim et al. in 2004 [16]. After that, graphene has become a hot spot in condensed matter physics due to its

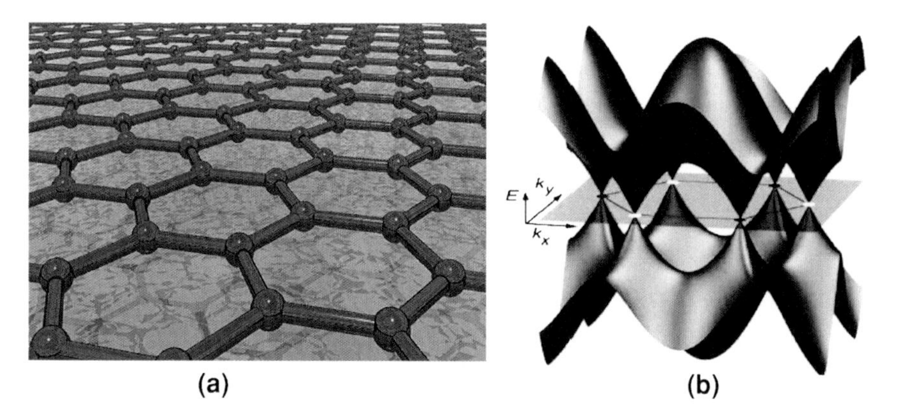

Fig. 4.4 Schematic presentation of **a** the lattice packing and **b** the band structure of a single layer graphene

Fig. 4.5 Structure of SLG and BLG. The *rectangular* unit cells are shown with *dashed lines* and the lattice vectors with *arrows*

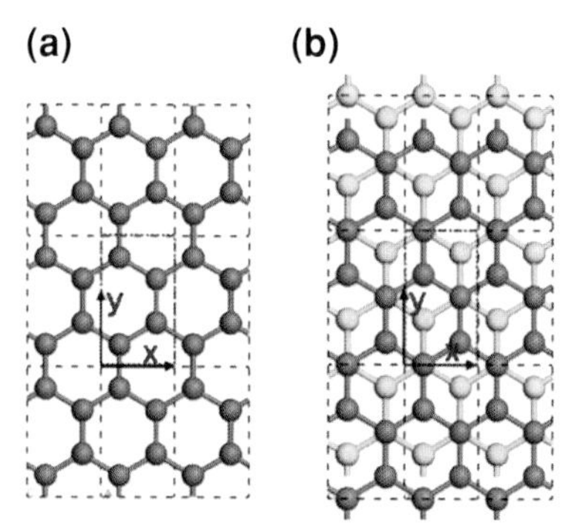

special electrical features [17]. As shown in Fig. 4.4b, the valence band and the conduction band only intersect at the K-point (also called the Dirac point). At this point, the energy dispersion is linear, which indicates that the effective mass of electrons is about zero, presenting a relativistic effect. The mobilities can exceed 15,000 cm^2/Vs even under ambient conditions [16]. Quantum hall effect (QHE) can be observed in graphene even at room temperature, extending the previous temperature range for QHE by a factor of 10 [18]. All these make graphene a candidate for new transport materials. Here, we study the charge transport in graphene sheets (GSs) and Graphene nanoribbons (GNRs).

4.3.1 Graphene Sheet

We investigate two kinds of graphene sheets, i.e., single layer graphene (SLG) and bi-layer graphene (BLG), as shown in Fig. 4.5. The geometry optimizations and band structure calculations are performed within VASP [12–14] with PBE

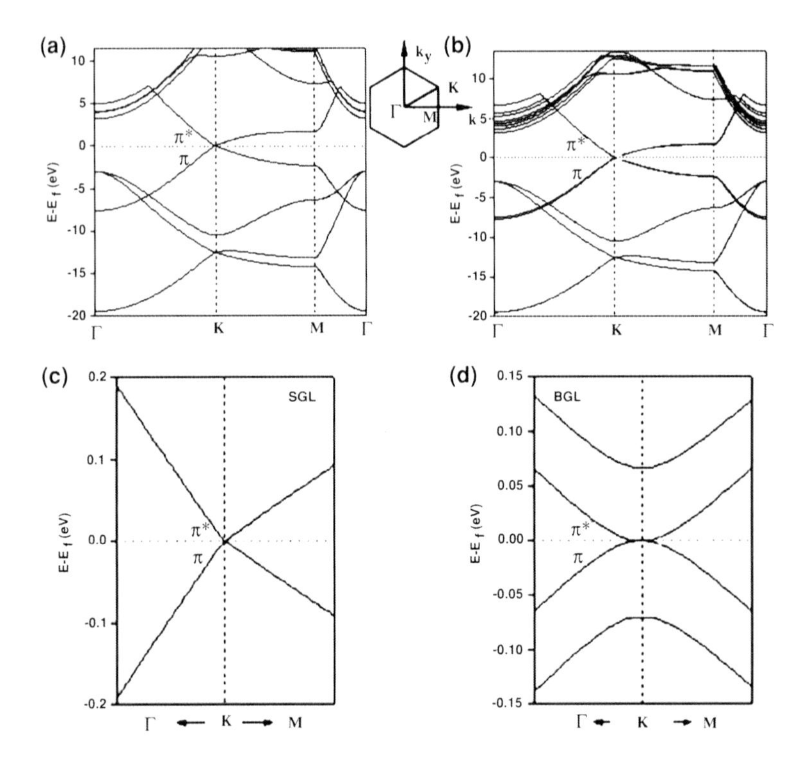

Fig. 4.6 Band structure of **a** SLG and **b** BLG. The fine structures near the Fermi level (expressed with *red dashed lines*) are replotted in **c** and **d**, respectively. The high symmetric points are shown together with the Brillouin zone

exchange correlation functional [19]. Vacuum layer thickness is set to be 30Å. In Fig. 4.6, we can find that the band structure of SLG and BLG are similar. The frontier π and π^* states intersect at K-point at the Fermi level, therefore both SLG and BLG are gapless semiconductors with a pseudo-metallic property. When comparing the band structure near the Fermi level, we find that the energy levels of π and π^* states are split from SLG to BLG. We first calculate the mobility explicitly according to Eq. 4.27. All the relevant results are presented in Table 4.3. In SLG, electron and hole mobilities are very close, while the difference is more significant in BLG. This is most probably due to the fact that the interaction between the two sheets influences the symmetry of π and π^* energy band structure. The deformation potential constants in both systems are more or less the same, however, the elastic constant of SLG is only half of BLG. Thus the mobility in BLG is generally larger than that of SLG, which indicates that the charge transport is benefitted from the double layer structure. We further apply the effective mass approximation. As listed in Table 4.4, the mobility is about 2.0–2.7 times of that

Table 4.3 Deformation potential constant E_1 [eV], elastic constant C [J/m^2], mobility μ [10^5cm^2/Vs] and relaxation time τ [ps] for SLG and BLG along x and y directions for both electron (π^*) and hole (π) transport

System	Band	E_1	C	μ	τ
SLG_x	π	5.140	328.019	3.217	13.804
	π^*			3.389	13.938
SLG_y	π	5.004	328.296	3.512	13.094
	π^*			3.202	13.221
BLG_x	π	5.330	680.167	3.949	16.158
	π^*			4.484	17.966
BLG_y	π	5.334	681.172	4.178	16.206
	π^*			4.636	18.019

Table 4.4 Effective mass fitted near the K-point m^* (the unit is 0.01 m_e, mass of an electron), and the corresponding mobilities μ^* [10^5cm^2/Vs] and relaxation times τ^* [ps]

System	Band	m^*	μ^*	τ^*
SLG_x	π	1.594	6.971	6.303
	π^*	1.597	6.914	6.277
SLG_y	π	1.594	6.872	6.431
	π^*	1.597	7.301	6.629
BLG_x	π	1.753	11.103	11.046
	π^*	1.750	11.105	11.047
BLG_y	π	1.753	11.103	11.046
	π^*	1.750	11.105	11.047

calculated directly from Boltzmann transport equation. However, the ratio of mobility between BLG and SLG remains the same. The result validates that the effective mass approximation can give reasonably good mobility values and nice relative trends.

4.3.2 Graphene Nanoribbon

Graphene can be cut into one-dimension nanoribbons in two different ways, namely, armchair graphene nanoribbons (AGNRs) and zigzag graphene nanoribbons (ZGNRs), as shown in Fig. 4.7. We are mostly interested in the influence of ribbon width, i.e., the number of carbon atoms along the side edge, N, on charge transport in both cases.

4.3.2.1 Armchaired Graphene Nanoribbon

Geometry optimization and band structure for AGNRs are computed at the same level as for the graphene sheets. The calculated energy band structures with

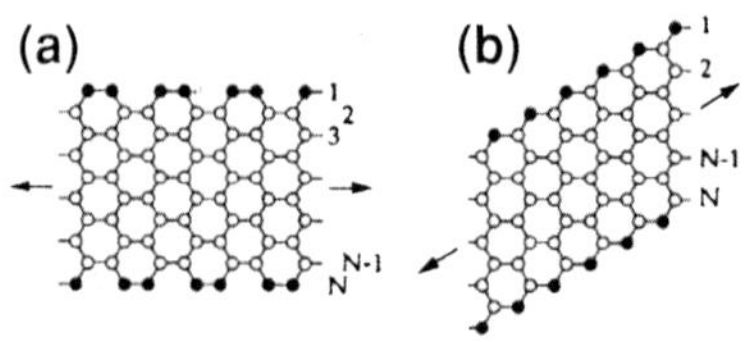

Fig. 4.7 Structure for graphene nanoribbons: **a** armchair and **b** zigzag. The ribbon width is N. The *top* and *bottom* edges were passivated by hydrogens

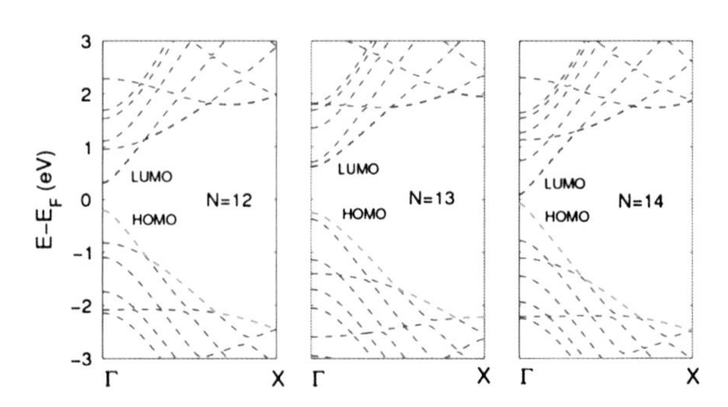

Fig. 4.8 Band structure of AGNRs (N = 12, 13, 14). Reprinted with permission for Ref. [20]. Copyright 2009 American Chemical Society

different widths are shown in Fig. 4.8. All investigated AGNRs are semiconductors due to the existence of the band gap. In Table 4.5, we show the parameters for AGNR with different widths which are realized with the method introduced in Sect. 4.1.7 using effective mass approximation. Electron and hole possess very close effective mass in the range 0.057–0.077m_e, where m_e is the charge of an electron. The elastic constant increases steadily with the ribbon width of AGNRs because the rigidity is enhanced. The deformation potential constant has a significant width dependence: for N = 3k, E_1 for holes is about one order of magnitude larger than that for electrons, while for N = 3k + 1 and 3k + 2, the situation is the opposite. To understand this width dependence, we examine the frontier molecular orbitals (see Fig. 4.9). For N = 12 (3k), it is noted that the bonding direction for the HOMO is perpendicular to the ribbon direction, while for the LUMO, the bonding direction is along the stretching axis, and thus HOMO is scattered more strongly by acoustic phonons than LUMO. For N = 3k + 1 and N = 3k + 2, the electron distribution of HOMO (LUMO) is close to the LUMO (HOMO) for N = 3k, and thus the deformation potential constant is larger for electron than hole. The periodicity and the magnitude change of the deformation potential constant result in an oscillating behavior in the width-dependent mobility with two orders of magnitude difference in the mobilities of electrons and holes (see Fig. 4.10) [20].

4.3.2.2 Zigzag Graphene Nanoribbon

The typical band structure of ZGNR is shown in Fig. 4.11. We notice that the conduction and valence bands merge flatly near the Fermi surface. Thus the effective mass approximation is not suitable any more. Instead we use Eq. 4.27 to study the width-dependent charge mobility (see Fig. 4.12). We can find that the

Table 4.5 Calculated width of ribbon W (including the passivated hydrogen), effective mass m^*, deformational potential constant E_1, the elastic constants C, the electron and hole mobility μ, and the average value of scattering time τ for AGNRs for $N = 9–17$, 33–35 and 42–44

N	W [nm]	Type	m^* [0.01 m_e]	E_1 [eV]	C [10^{11} eV/cm]	μ [10^4cm^2/Vs]	τ [ps]
9	1.176	e	7.21	1.11	3.24	108.11	44.33
		h	6.04	11.00		1.44	0.49
10	1.300	e	7.87	10.992	3.59	1.07	0.48
		h	5.71	2.47		34.30	11.14
11	1.420	e	6.79	10.838	3.81	1.46	0.56
		h	6.51	1.904		50.34	18.63
12	1.543	e	7.17	1.230	4.29	117.36	47.85
		h	6.26	10.904		1.83	0.65
13	1.681	e	7.68	10.972	4.64	1.44	0.63
		h	6.00	2.32		46.71	15.92
14	1.792	e	6.88	10.892	4.84	1.80	0.70
		h	6.65	1.870		64.14	24.26
15	1.913	e	7.23	1.312	5.35	127.26	52.29
		h	6.43	10.960		2.17	0.79
16	2.036	e	7.63	10.954	5.70	1.79	0.78
		h	6.23	2.21		59.74	21.15
17	2.156	e	6.99	11.158	5.87	2.03	0.81
		h	6.81	1.842		77.35	29.96
33	41.19	e	7.02	1.77	12.70	172.84	69.02
		h	6.68	19.2		4.91	1.86
34	42.39	e	7.17	10.84	12.80	4.52	1.84
		h	6.61	2.34		110.07	41.32
35	43.60	e	7.03	18.9	12.98	4.80	1.91
		h	6.88	1.91		130.12	51.8
42	52.24	e	6.94	1.81	17.11	212.16	90.66
		h	6.69	11.19		6.28	2.39
43	53.48	e	7.07	10.96	17.34	6.11	2.46
		h	6.62	2.21		165.95	62.42
44	54.79	e	7.06	10.80	17.45	6.21	2.50
		h	6.70	2.19		185.04	70.8

mobility in ZGNRs is about two orders of magnitude lower than that of the AGNRs, which is due to the flat band near the Fermi surface suggesting a larger effective mass [20]. Overall, hole mobility is several times larger than electron, and there is no width-dependent carrier polarity as we have observed in AGNRs.

4.4 Application: Graphdiyne

Graphdiyne is one of the most "synthetically approachable" allotropes containing two acetylenic (diacetylenic) linkages between carbon hexagons [21]. It has been predicted to exhibit fascinating properties including extreme hardness, good

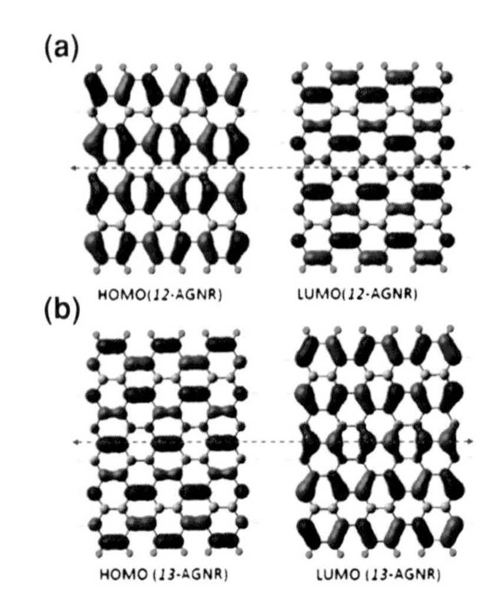

Fig. 4.9 Γ-point HOMO and LUMO wave functions for AGNRs with **a** $N = 12$ and **b** $N = 13$. The *red dashed line* stands for the direction of stretching. Reprinted with permission for Ref. [20]. Copyright 2009 American Chemical Society

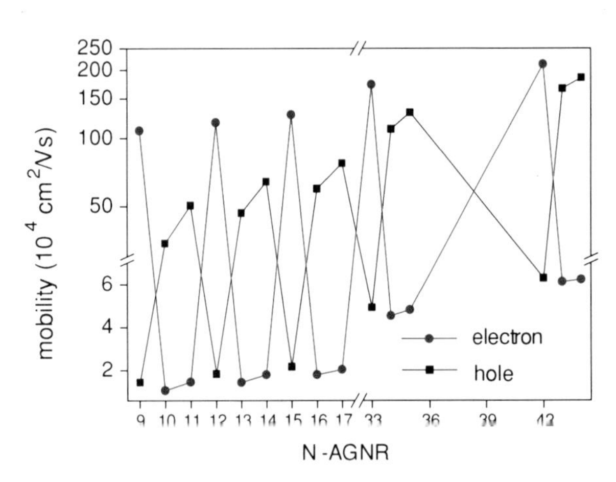

Fig. 4.10 Theoretically predicted mobility dependence on the width of AGNRs for both electron and hole. Reprinted with permission for Ref. [20]. Copyright 2009 American Chemical Society

stability, large third-order nonlinear optical susceptibility, high fluorescence efficiency, high thermal resistance, nice conductivity or superconductivity and through-sheet transport of ions [21–23]. In this section, we investigate the electronic structure and charge transport of graphdiyne sheet (GDS) and its various graphdiyne nanoribbons (GDNRs) through first principles calculations.

4.4.1 Graphdiyne Sheet

The structure of a graphdiyne sheet is shown in Fig. 4.13. VASP [12–14] is used to optimize the lattice and calculate the energy band at the DFT/LDA level [24]. In Fig. 4.14, a band gap of 0.46 eV is found at the Γ-point which means that the

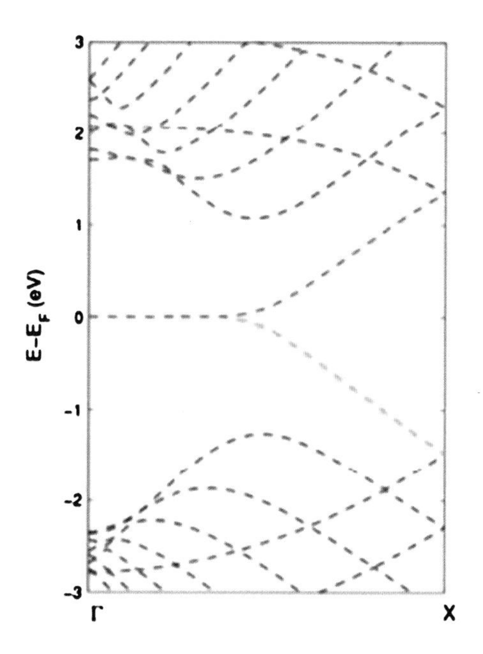

Fig. 4.11 Band structure for ZGNR ($N = 8$). Reprinted with permission for Ref. [20]. Copyright 2009 American Chemical Society

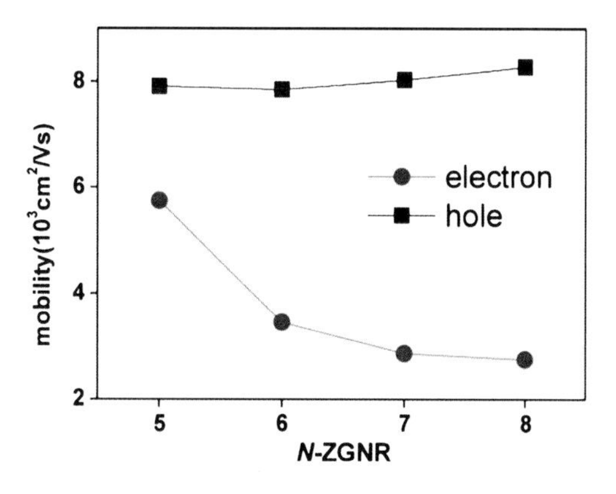

Fig. 4.12 Calculated width-dependent mobility for ZGNR. Reprinted with permission for Ref. [20]. Copyright 2009 American Chemical Society

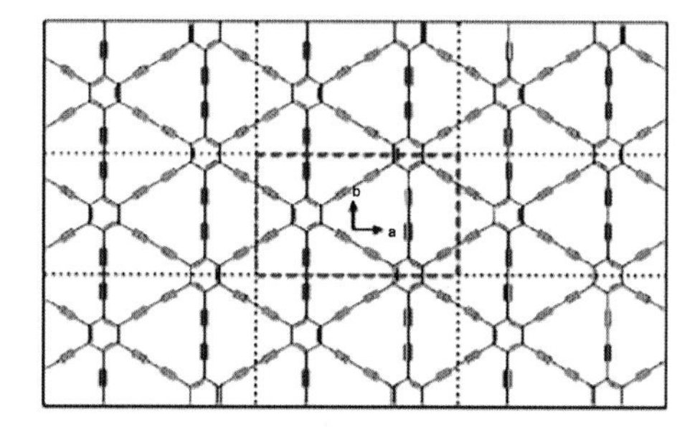

Fig. 4.13 Schematic representation of a single graphdiyne sheet. The *rectangular* supercell is drawn with *dashed lines* and the lattice vectors are shown as *arrows*. Reprinted with permission for Ref. [24]. Copyright 2011 American Chemical Society

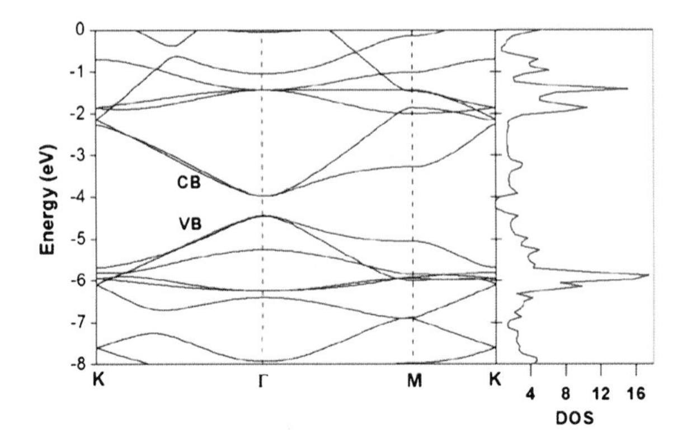

Fig. 4.14 DFT-calculated band structure and density of states for a single graphdiyne sheet. The Brillouin zone with the chosen high symmetric k-points is also shown. Reprinted with permission for Ref. [24]. Copyright 2011 American Chemical Society

Table 4.6 Deformation potential E_1, elastic constant C, carrier mobility μ and the averaged value of scattering relaxation time τ at 300 K for electrons and holes along a and b directions in a single graphdiyne sheet

Carrier type	E_1 [eV]	C [J/m^2]	μ [10^4cm^2/Vs]	τ [ps]
e^a	2.09	158.57	20.81	19.11
h^a	6.30	158.57	1.97	1.94
e^b	2.19	144.90	17.22	15.87
h^b	6.11	144.90	1.91	1.88

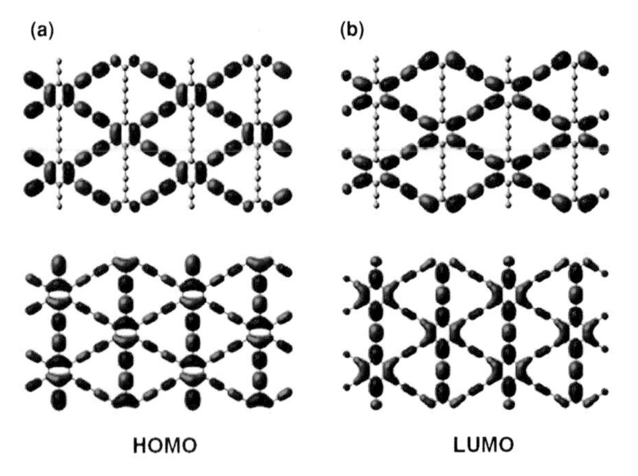

Fig. 4.15 Γ-point degenerate HOMO and LUMO density distributions for a graphdiyne sheet. Reprinted with permission for Ref. [24]. Copyright 2011 American Chemical Society

graphdiyne sheet is a semiconductor. The deformation potential constants and elastic constants are obtained through the fitting approach described in Sect. 4.1.7. The mobility and relaxation time are calculated through Eqs. 4.27 and 4.21, respectively, without using the effective mass approximation. The results are listed in Table 4.6. We notice that the in-plane mobilities along directions a and b are close to each other for both electron and hole. Besides, the room temperature electron mobility is about

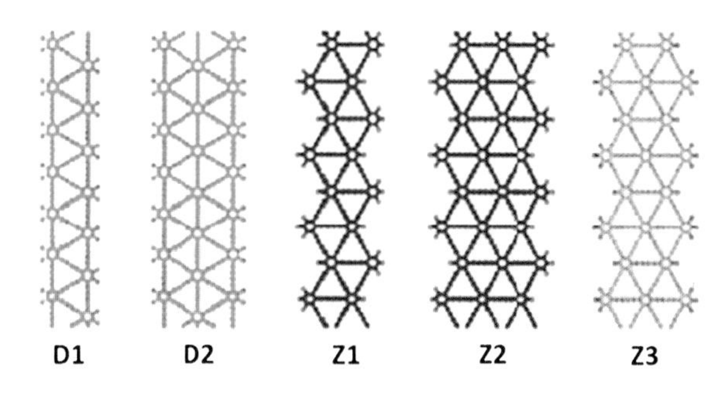

Fig. 4.16 Schematic representation of five different GDNRs. D1 and D2 are divan GDNRS (denoted as DGDNR) with two and three carbon hexagons in width, respectively. Z1, Z2 and Z3 are zigzag GDNRs (denoted as ZGDNR) with two, three and alternating width, respectively. Reprinted with permission for Ref. [24]. Copyright 2011 American Chemical Society

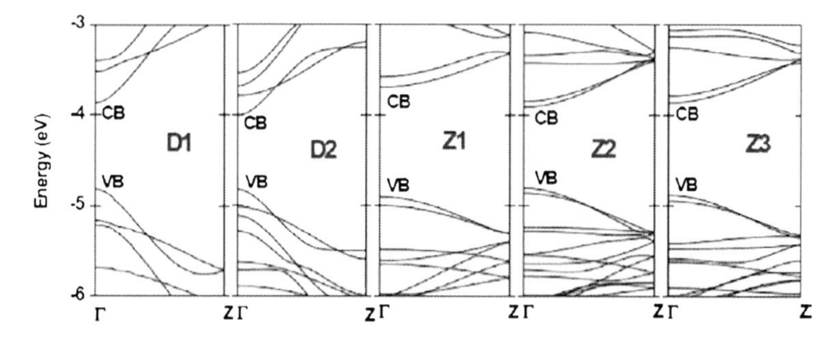

Fig. 4.17 Band structures of the investigated five GDNRs. Reprinted with permission for Ref. [24]. Copyright 2011 American Chemical Society

one order of magnitude higher than that of hole. Such difference in electron and hole mobilities can be attributed to the deformation potential E_1: E_1 for hole is three times as large as that for electron. E_1 is a characterization of the coupling strength of electron or hole to acoustic phonons, and can be understood by checking the frontier molecular orbitals responsible for transport. The highest occupied molecular orbital (HOMO) and the lowest unoccupied molecular orbital (LUMO) at Γ-point are shown in Fig. 4.15. We can see that HOMO exhibits anti-bonding character between carbon hexagons and diacetylenic linkages, whereas LUMO exhibits bonding feature. As a result there are more nodes in HOMO than LUMO in either direction *a* or direction b. And thus the site energy for hole is more prone to change, and its deformation potential is larger.

4.4.2 Graphdiyne Nanoribbons

Using graphdiyne sheet as a template, five different GDNRs are chosen, as shown in Fig. 4.16. D1 and D2 are DGDNRs, while Z1, Z2 and Z3 are ZGDNRs with

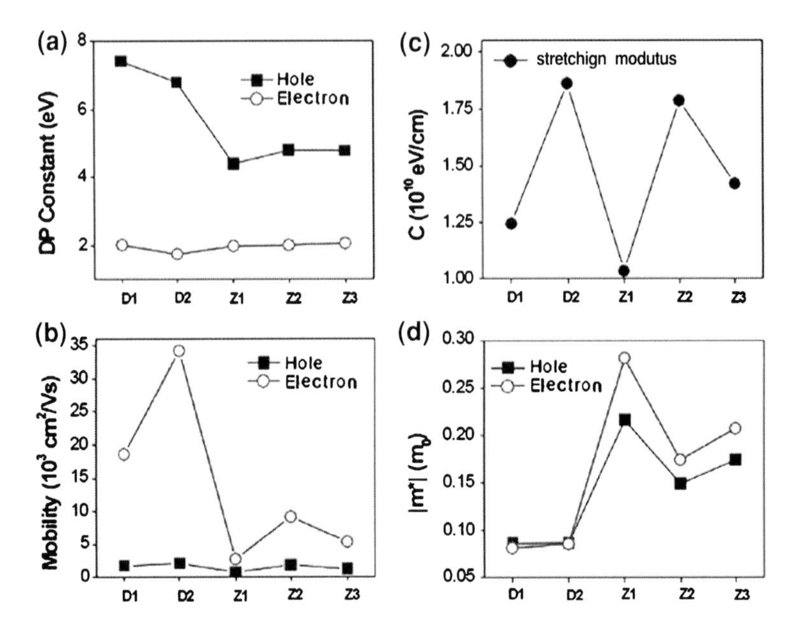

Fig. 4.18 **a** Deformation potential constants, **b** charge mobility, **c** elastic constants, and **d** effective masses for holes and electrons in five GDNRs. Reprinted with permission for Ref. [24]. Copyright 2011 American Chemical Society

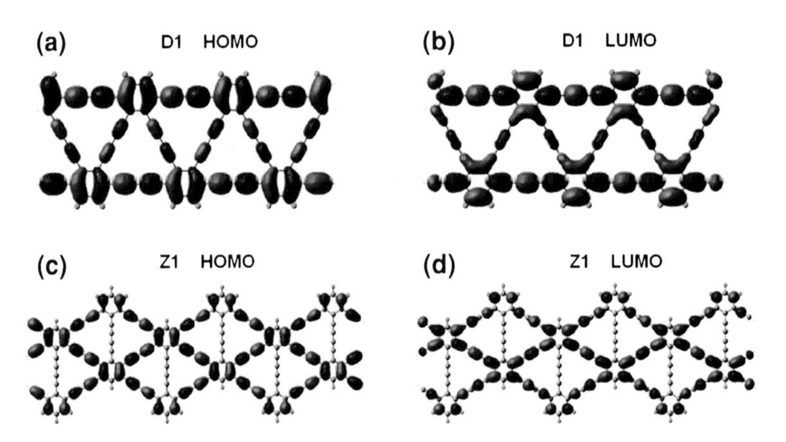

Fig. 4.19 Γ-point HOMO and LUMO electronic density distributions for GDNRs D1 and Z1. Reprinted with permission for Ref. [24]. Copyright 2011 American Chemical Society

different widths. The DFT-calculated band structures are shown in Fig. 4.17. All five nanoribbons are predicted to be semiconductors. The smallest band gap of about 0.8 eV is found in D2. The deformation potential constants, elastic constants, mobilities and effective masses are calculated. From Fig. 4.18a, we

Table 4.7 Calculated band gap, effective mass (m_h^* and m_e^*), deformation potential constants for VB and CB (E_V and E_C), elastic constant C, carrier mobility μ at 300 K for five GDNRs.

	D1	D2	Z1	Z2	Z3
Band gap [eV]	0.954	0.817	1.205	0.895	1.015
m_h* [m0]	0.086	0.087	0.216	0.149	0.174
m_e* [m0]	0.081	0.086	0.281	0.174	0.207
E_V [eV]	7.406	6.790	4.386	4.786	4.776
E_C [eV]	2.006	1.730	1.972	2.000	2.054
C [10^{10} eV/cm]	1.244	1.864	1.035	1.787	1.420
μ_h [10^3 cm^2/Vs]	1.696	2.088	0.755	1.815	1.194
μ_e [10^3 cm^2/Vs]	18.590	34.241	2.692	9.127	5.329
μ_h* [10^3 cm^2/Vs]	0.711	1.253	0.426	1.073	0.679
μ_e* [10^3 cm^2/Vs]	10.580	19.731	1.418	5.015	2.829

Note that μ_h and μ_e are calculated without effective mass approximation, while μ_h^* and μ_e^* are calculated with effective mass approximation

notice that E_1 for hole is much larger than electron for all the GDNRs, agreeing with the graphdiyne sheet. The same antibonding feature between carbon hexagons and diacetylenic linkages has been found for the HOMO, whereas the bonding feature is found for the LUMO (see Fig. 4.19). As a result, the intrinsic electron mobility is significantly larger than that of hole. It is seen that the DGDNRs is more favorable than the ZGDNRs for electron transport since the LUMO for DGDNRs is much more delocalized in the direction of ribbon axis than that for ZGDNRs (see Fig. 4.19). Besides, the charge mobility increases with width within the same class of GDNRs. We also adopt the effective mass approximation since the mobility formula only has three parameters, i.e., the deformation potential constant, the elastic constant and the effective mass, which are quite helpful to better understand the observed transport phenomena. The results are shown in Table 4.7. The results using the effective mass approximation are in good agreement with that calculated without effective mass approximation. We find that the elastic constant increases with the width in the same class of GDNRs as shown in Fig. 4.18c, thereby the mobility has the same behavior. And the effective masses of DGNRs are smaller than those of ZGNRs, suggesting that DGNRs should have larger mobility than ZGNRs.

4.5 Conclusions

In this chapter, we have introduced the basic concepts of the deformation potential theory based on the Boltzmann transport equation and the effective mass approximation. We have applied this approach to oligoacences, graphene and graphdiyne sheets and nanoribbons. We found that (1) the scattering intensity of charge with acoustic phonons is about three times as large as that with optic phonons in molecular crystals; (2) the obtained temperature dependence of mobility follows the power law; (3) the width of the ribbon plays an important role in tuning the polarity of carrier transport in armchair graphene nanoribbons, while there is no alternating size dependence for zigzag graphene nanoribbons; (4) electron mobility is very high in both single graphdiyne sheets and graphdiyne nanoribbons and the charge mobility increases with the width of nanoribbon. We note that the present approach considers only the acoustic scatterings and needs to be improved. To fully understand the role of electron–phonon couplings on charge transport, optical phonons should be considered together with the acoustic phonons at the same level of theoretical treatment. Accurate numerical simulations are needed for benchmark studies on general models with wide parameter range.

References

1. J. Bardeen, W. Shockley, Phys. Rev. **80**, 72 (1950)
2. Y.C. Cheng, R.J. Silbey, D.A. da Silva Filho, J.P. Calbert, J. Cornil, J.-L. Brédas, J. Chem. Phys. **118**, 3764 (2003)
3. L. Tang, M.Q. Long, D. Wang, Z. Shuai, Sci. China Ser. B-Chem. **52**, 1646 (2009)
4. F.B. Beleznay, F. Bogar, J. Ladik, J. Chem. Phys. **119**, 5690 (2003)
5. N. Karl, in *Landolt-Börnstein: Numerical Data and Functional Relationship in Science and Technology*, Group III, vol. 17. ed. by K.-H. Hellwege, O. Madelung (Springer, Berlin, 1985), p. 106
6. K. Hannewald, P.A. Bobbert, Appl. Phys. Lett. **85**, 1535 (2004)
7. G.K.H. Madsen, D.J. Singh, Comput. Phys. Commun. **175**, 67 (2006)
8. S.H. Wei, A. Zunger, Phys. Rev. B **60**, 5404 (1999)
9. V.I. Ponomarev, O.S. Filipenko, L.O. Atovmyan, Kristallografiya **21**, 392 (1976)
10. C.P. Brock, J.D. Dunitz, Acta. Cryst. B **46**, 795 (1990)
11. D. Holmes, S. Kumaraswamy, A.J. Matzger, K.P.C. Vollhardt, Chem. Eur. J. **5**, 3399 (1999)
12. G. Kresse, J. Hafner, Phys. Rev. B **47**, 558 (1993)
13. G. Kresse, J. Hafner, Phys. Rev. B **49**, 14251 (1994)
14. G. Kresse, J. Furthmüller, Phys. Rev. B **54**, 11169 (1996)
15. L.J. Wang, Q. Peng, Q.K. Li, Z. Shuai, J. Chem. Phys. **127**, 044506 (2007)
16. K.S. Novoselov, A.K. Geim, S.V. Morozov, D. Jiang, Y. Zhang, S.V. Dubonos, I.V. Grigorieva, A.A. Firsov, Science **306**, 666 (2004)
17. A.K. Geim, K.S. Novoselov, Nat. Mater.**6**, 183 (2007)
18. Y. Zhang, J.W. Tan, H.L. Stormer, P. Kim, Nature **438**, 201 (2005)
19. J.P. Perdew, K. Burke, M. Ernzerhof, Phys. Rev. Lett. **77**, 3865 (1996)
20. M.Q. Long, L. Tang, D. Wang, L.J. Wang, Z. Shuai, J. Am. Chem. Sci. **131**, 17728 (2009)
21. M.M. Haley, W.B. Wan, Adv. Strained Interesting Org. Mol. **8**, 1 (2000)
22. R.H. Baughman, H. Eckhardt, M. Kertesz, J. Chem. Phys. **87**, 6687 (1987)
23. A.T. Balaban, C.C. Rentia, E. Ciupitu, Rev. Roum. Chim. **13**, 231 (1968)
24. M.Q. Long, L. Tang, D. Wang, Y.L. Li, Z. Shuai, ACS Nano **5**, 2593 (2011)

Chapter 5
Outlook

Modeling the charge transport in organic materials is a formidable task due to the complexity in dealing with various scattering mechanisms, which is intrinsically a many-body problem [1]. In this book, we present three approaches, namely, the localized hopping model, the extended band model, and the polaron model, to compute the mobility for organic and carbon materials at the first-principles level. We show that we indeed achieved some successes in, for instance, predicting the intrinsic mobility values from given materials structures, or in rationalizing the structure–property relationship. However, the story is far from complete. More complicated treatments for the many-body effects remain a challenging issue. At present, the picture is still fragmented in the sense that the hopping and the polaron models consider only the optical phonons, while the band model only considers the acoustic phonons under the deformation potential approximation. Therefore, better descriptions rely on combining both phonon scatterings as well as the inclusion of phonon dispersion. Besides, to go beyond the assumption that the charge is localized or delocalized, direct quantum dynamics [2] and/or mixed quantum–classical dynamics [3] can also be used. Through solving the time-dependent Schrödinger equation or the time-dependent Liouville equation for electron, one can capture the essence of charge transport in electron–phonon interacting systems. Due to the expensive computational cost, current studies are limited to one-dimensional molecular arrays with very few phonon modes. However, recent studies confirmed that the feedback from electron dynamics to nuclear vibrations can be fully neglected for molecular crystals with large mobility [4]. This approach is quite promising for obtaining the absolute magnitude of the charge mobility for organic materials, because the charge dynamics can be modeled "on the fly" through a hybrid approach combining molecular dynamics simulations of nuclear motion, quantum-chemical calculations of the electronic Hamiltonian at each geometric configuration, and time-dependent electron dynamics. We expect that these numerical approaches, as well as the three models described in this book,

Z. Shuai et al., *Theory of Charge Transport in Carbon Electronic Materials*,
SpringerBriefs in Molecular Science, DOI: 10.1007/978-3-642-25076-7_5,

could be further developed and benchmarked at consistent parameter level, to get a full understanding of the charge transport mechanism.

The crystal structure is indispensable to computational predictions of the intrinsic charge mobility. Generally, this structure is taken from experimental measurements, making computations impractical for new molecules when there is no knowledge of the crystal packing structures. Besides, morphological information is necessary to study the charge transport for less ordered realistic systems. Thus it is highly desirable to develop computational methods to predict the crystal structures and long range geometries, to meet the demands of molecular design in organic electronics [5].

There is no doubt that, driven by the remarkable advances in materials and devices and higher capacity of computational technique, our understanding of charge transport is moving rapidly toward a quantitative description of charge mobility through comprehensive consideration of electron–phonon scatterings and more realistic molecular packings. It is a formidable and necessary task to get quantitative description for charge transport in organic electronic materials, in order to help materials design. There is still a long way to go.

References

1. Z.G. Shuai, L.J. Wang, Q.K. Li, Adv. Mater. **23**, 1145 (2011)
2. D. Wang, L.P. Chen, R.H. Zheng, L.J. Wang, Q. Shi, J. Chem. Phys. **132**, 081101 (2010)
3. A. Troisi, G. Orlandi, Phys. Rev. Lett. **96**, 086601 (2006)
4. L.J. Wang, D. Beljonne, L.P. Chen, Q. Shi, J. Chem. Phys. **134**, 244116 (2011)
5. G.M. Day, T.G. Cooper, A.J. Cruz-Cabeza, K.E. Hejczyk, H.L. Ammon, S.X.M. Boerrigter, J.S. Tan, R.G. Della Valle, E. Venuti, J. Jose, S.R. Gadre, G.R. Desiraju, T.S. Thakur, B.P. van Eijck, J.C. Facelli, V.E. Bazterra, M.B. Ferraro, D.W.M. Hofmann, M.A. Neumann, F.J.J. Leusen, J. Kendrick, S.L. Price, A.J. Misquitta, P.G. Karamertzanis, G.W.A. Welch, H.A. Scheraga, Y.A. Arnautova, M.U. Schmidt, J. van de Streek, A.K. Wolf, B. Schweizer, Acta Crystallogr. Sect. B Struct. Sci. **65**, 107 (2009)

Printed by Publishers' Graphics LLC
CAMZ130908.20.05.101